Andrea JoHo BAKER

An Introduction to Urban
Geographic Information Systems

Spatial Information Systems

General Editors

P.H.T. BECKETT

M. GOODCHILD

P. A. BURROUGH

P. SWITZER

An Introduction to Urban Geographic Information Systems

WILLIAM E. HUXHOLD

University of Wisconsin—Milwaukee

New York   Oxford
OXFORD UNIVERSITY PRESS
1991

Oxford University Press

Oxford   New York   Toronto
Delhi   Bombay   Calcutta   Madras   Karachi
Petaling Jaya   Singapore   Hong Kong   Tokyo
Nairobi   Dar es Salaam   Cape Town
Melbourne   Auckland

and associated companies in
Berlin   Ibadan

Copyright © 1991 by Oxford University Press, Inc.

Published by Oxford University Press, Inc.,
200 Madison Avenue, New York, New York 10016

Library of Congress Cataloging-in-Publication Data
Huxhold, William E.
An introduction to urban geographic information systems
William E. Huxhold.
p. cm.   Includes bibliographical references.
ISBN 0-19-506534-4.
ISBN 0-19-506535-2 (pbk.)
1. Geography—Data processing.
2. Cities and towns—Data processing.   I. Title.
G70.2.H88   1991   910'.285—dc20

1 3 5 7 9 8 6 4 2

Printed in the United States of America
on acid-free paper

This book is dedicated to my father,
Earle Huxhold

Spatial Information Systems

The collation of data about the spatial distribution of significant properties of the earth's surface and of people, animals, and plants has long been an important part of the activities of organized societies. Until relatively recently, however, most of these data were kept in the form of paper documents and maps from which they could be read off easily, but only with difficulty could they be used to analyse the patterns of distribution of attributes over the earth's surface and the processes that had given rise to them. The developments in both computer technology and mathematical tools for spatial analysis that have taken place in the second half of the twentieth century have made many things possible, among them the abilities to store, to retrieve at will, singly or in combination, and to display data about all aspects of the earth's surface. As well as being able to handle existing data, the latest spatial information systems can just as easily handle fictional data and the results of simulation models, permitting scenarios of possible past or future situations to be modelled and explored. These abilities have created a revolution in the mapping sciences and in their uses in the practical day-to-day inventory, understanding, and management of our environment. Today, computerized spatial information systems are used in many branches of pure and applied science, in business and commerce, and in local, national, and international governmental agencies. The applications range from the completely utilitarian, such as mapping the networks for telephones, electricity, and sewers, to the esoteric and futuristic, such as in modelling the possible future effects of climatic change.

The rapid growth in interest in spatial information systems and the accelerating rate at which they are being used has outstripped the normal rate of supply of trained scientists and technicians in the mapping and environmental sciences. Universities and polytechnics the world over are beginning to set up courses on spatial analysis and geographical information systems, not only to educate their own students but also to reeducate staff from government and business. Therefore, one aim of this series is to provide the basic source texts for the courses for third-year undergraduates, post-graduates, and practitioners, especially interdisciplinary texts covering basic principles that can be applied in many fields.

At present new knowledge about the theory and practice of modern spatial information systems is only available from conference proceedings and scientific articles published in a limited range of books and journals. The second aim of this series is to provide scholarly monographs, written by experts, in order to bring order and structure into this rapdily developing field. These monographs will gather new knowledge from diverse sources and present it to researchers and practitioners so that it becomes widely understandable and available.

One of the most striking aspects of computerized spatial information systems is that, when they are used to bring data about many different kinds of spatial patterns together, the results rarely fail to surprise and delight. There is, of course, a danger that the user will be too easily impressed by sheer technology. It is the aim of this series to ensure that users of spatial information systems are not only impressed by technology, but are also really delighted because they have achieved a deeper understanding of the world around them.

P. H. T. Beckett *(editor)*
Department of Agricultural Science
University of Oxford

P. A. Burrough *(editor in chief)*
Professor of Geography
University of Utrecht

M. Goodchild *(editor)*
Co-Director, National Center for Geographic Information and Analysis
University of California, Santa Barbara

P. Switzer *(editor)*
Professor of Statistics
Stanford University, California

Foreword

Jack Dangermond, *President*
Environmental Systems Research Institute

William E. Huxhold's book, *An Introduction to Urban Geographic Information Systems,* is one of the first of what I hope will be a growing number of books that describe the application of geographic information system (GIS) technology to particular fields and disciplines. While many general textbooks about GIS are now beginning to appear, and these are useful in educating people about GIS technology, real growth in the *use* of this technology depends chiefly on people who can find ways to put GIS's to practical use. Books such as this make a critical contribution to that process by helping people see exactly how GIS technology can benefit their work.

Huxhold, in his preface, says he intends to answer the questions "What is (a GIS), what can it do, and what is its value?"

In answering these questions, he begins by describing the fundamental value of urban geographic information, leads the reader through the definitions of an urban GIS, and then discusses various applications of such a system. Chapters 4 and 5 deal with the important technical topics of topological data and the value of a geographic base file to an urban GIS. The difficult practical problems of creating automated land records information systems are mentioned in Chapter 6, together with some of the technology being applied to their solution.

The final chapter, "The Model Urban GIS Project," provides an especially important service to the reader; it explains in some detail just how to go about *implementing* a GIS in an urban setting, from initial idea through final working system. It discusses the practical matters of obtaining organizational and financial support, identifies the pitfalls of the process, and stresses such practical techniques as the pilot study, to help ensure the project's success.

Topics especially relevant to urban spatial information systems, which are usually only "mentioned" in other books, receive a more generous treatment here. The sections dealing with the Census Bureau's TIGER files and the NAVSTAR global positioning system (GPS) are good examples of this, and are particularly timely; they provide sufficient information for the reader to quickly obtain a good grasp of these important subjects.

The book should be useful and valuable to a variety of potential readers. It is particularly directed to persons who are already working with urban information, but who may have little or no knowledge of geographic information system (GIS) technology. These readers will find much in the book that is already familiar to them in the situations, problems, and many examples of urban information use that have been included in the text; but they will be led to see this familiar material in a new light, because Huxhold's aim is to show, in some detail, just how GIS technology can assist users of urban information in their work. The book is full of specific examples of urban GIS applications, supported by anecdotes, actual examples, illustrations of GIS output, and by a coherent series of practical exercises. There is a good deal of detail in this book, but it has been chosen with the urban information user clearly in mind. The urban examples are numerous, practical, and drawn from extensive experience in cities and counties, especially in the Milwaukee, Wisconsin area, with which Huxhold is intimately familiar.

This is also a good book for persons who must manage urban GIS's. Few books in the GIS field provide real help to GIS managers. This one provides a broad overview, a sound outline of the technology as applied to urban settings, practical help in understanding how an urban GIS works, and a good deal of useful wisdom, offered in the form of advice and actual experience.

This material will probably be extremely useful to persons unfamiliar with urban information systems and urban GIS's, because they will find an overview difficult to obtain from the rather scattered literature of this field. Readers can also profit immediately from the considerable practical experience of the author in dealing with urban problems through the use of geographic information.

The book is certainly suitable as a textbook for a course in urban GIS technology; it contains the necessary definitions, explanations, illustrations, examples, and practical exercises that will help students grasp, understand, and apply the subject. The exercises are designed to give the reader a "feel" for what a GIS can do *without* requiring that the reader have access to a GIS as the book is read; the exercises could also be adapted to a course in which a GIS was available for student use. Huxhold's experience in teaching courses of these kinds, over a number of years, is evident in his approach, organization, and selection of materials.

The book can be read as a quick overview of the field, or studied at length, since there is ample substance to provoke thought and reflection. Enough references are provided so that the text is not burdened, but the reader wanting more information can identify many of the important literature resources available.

Beyond all these points, however, the book can also be recommended for its clear and direct style, the thought that has gone into the selection of topics, and the author's obvious interest in helping readers learn about and make use of GIS technology in an urban setting.

GIS technology is especially useful in providing us with a more complete picture of the *relationships* between all the elements of the various natural and cultural systems on which we depend. Huxhold understands and appreciates these relationships in an urban setting, and makes an important contribution to helping the rest of us see urban geographic information systems as he does.

Preface

On Saturday morning, July 19, 1987, the following message was broadcast on the National Public Radio network to homes all across the nation:

GIS, geographical information systems research, is the hottest thing in geography and map-making these days . . . and the power of GIS has really nothing to do with simply translating paper maps into video screen pictures. It is the prospect of combining map images with other kinds of information—pictures from satellites, statistics from census data—to be able to display relationships graphically that were impossible before the computer.

Imagine, for instance, a map showing all of the grazing lands in the U.S.:

"Mooooo . . ."

with deposits of uranium:

"Click . . click . . ."

owned by people named Fred or Martha:

"Howdy, Fred. Mornin', Martha"

that are also located near a bus stop:

"Honk . . honk . . ."

With GIS, you can show all of this on the same map:

"Moooo . . Click . . click . . Howdy, Fred. Mornin', Martha . . Honk."

(It's also a heck of a lot more comprehensible than this audio simulation.)

Such a study might occupy a Ph.D. candidate or real estate agent for years. With GIS, it takes a few seconds. One of the challenges of the

scientists constructing this system is to be flexible enough to envision future uses—even ridiculous ones that can't be foreseen today.*

Thus, the nation (or at least those of us who were up that morning
and were listening to the broadcast) was introduced to geographic information systems in terms that were understandable to even the most
"computer phobic" of people: "combining map images with other kinds
of information." This is the essence of GIS.

Contributions from many different disciplines (geography, surveying, data processing, engineering, planning, environmental science,
landscape architecture, and others) have been necessary to develop the
power and build flexibility into geographic information systems technology. Because of the diversity of these disciplines, however, there
was for many years little agreement on what a GIS is and what it should
do. Each discipline seemed to have its own name for this new technology: computer graphics, computer-aided mapping, computer-assisted
cartography, automated mapping and facilities management, and others. Many tags have been attached to this new technology for processing digital map information and the data associated with locations on
that map. To some, the technology has provided a better way to produce maps; to others, it has allowed for the overlaying of different
maps on top of each other; and to yet others, the new technology has
been a new method for combining data from many different sources
(maps as well as tabular data) for the purpose of analyzing spatial relationships among data related to locations on the earth. GIS does it
all; it improves the way we use maps, and it improves the way we
analyze data about features located on the earth.

It was not until the 1970s that the phrase "geographic information
system" was first used to name a set of tools for creating, maintaining,
analyzing, and displaying maps and data for use in public agencies.
Prior to that, computer-aided mapping (CAM) was generally accepted
as the new technology for improving the process of map making. Engineers, surveyors, and draftsmen embraced the automated mapping
technology because it was a new method for making their design and
drafting tasks more efficient. (They could point to an unwanted line
with a cursor and it would disappear, or move it to a parallel location—
all without erasing and redrawing.) Design and drafting work could be
done faster, which meant more efficiency in government and lower
costs, translating into lower property taxes. On the private side, efficiency meant that more business could be created without adding costs—
resulting in higher profits and bigger dividends for the stockholders.

While the engineers, surveyors, and draftsmen were enjoying the

*This report was originally broadcast on National Public Radio's news and information
magazine "Weekend Edition" on July 19, 1987 and is printed with the permission of
National Public Radio. Any unauthorized duplication is prohibited.

efficiency of this new technology, the urban planners, geographers, environmental scientists, and landscape architects were using the new computer-assisted cartography technology to perform spatial analyses on geographic data. Using very large data bases, statistical data, and diverse sources of information, computer graphics became welcome, efficient tools for those complex calculations and geographic combinations needed to perform research and analyze data related to geography. Never mind the accuracy of the resulting display. (A census tract, an aldermanic district, or a soil boundary on a small-scale map needs very little accuracy on its map location.) As long as the entire geographic area of study was represented, what difference did the placement of a polygon boundary within ten to twenty feet of accuracy on the ground mean? (Take me out to the boulevard and show me where that census tract boundary is!) The accuracy and precision demanded by the engineers, surveyors, and draftsmen in their automated mapping systems were of little interest to the planners, geographers, and scientists who required capabilities to perform spatial analyses from this new technology. The engineers, surveyors, and draftsmen, on the other hand, were not spatial analysts, but were merely looking for a more efficient method for performing their highly accurate work.

Thus—the conflict between accuracy and analysis. Local government administrators, budget directors, data processors, and elected officials were caught between the two groups of professions when it came time to consider the costs and benefits of implementing the new technology. Both groups wanted automated mapping and spatial analysis, but why buy both when the technology is essentially the same? (Some jurisdictions, by the way, did choose both—Orange County, California, for one and Anchorage, Alaska, for another.)

This book cannot answer that question. Each local government has its own priorities, its own personalities, and its own resources influencing that decision. This book can, however, answer the question, "What is a geographic information system?" and readers who are caught between the two professional opinions will, hopefully, be able to choose which of the GIS technologies is important to their jurisdiction. Their questions answered in this book are: "What is it, what can it do, and what is its value?"

Students, who will eventually be the urban planners, the engineers, the data processors, administrators, and policy-makers of our urban future, must know what a geographic information system is, what it can do, and what its value is, because graduates will no doubt be confronted with these questions soon after receiving their first paychecks. Some must make GIS work in their municipalities or counties. Some must sell GIS products and services to these public agencies. And some must work together with local agencies to share in the costs and benefits of public/private cooperative GIS projects so that governments, utility companies, and other related enterprises can better serve the taxpayers, the rate payers, and the citizens of our urban environment.

Chapter 1 addresses the value of information and how information systems have evolved from improving efficiency to improving effectiveness in local government service delivery, management, and policy-planning activities. Information systems are presented in terms of the information needed to run government rather than the technology used to process it.

Chapter 2 defines the urban geographic information system as a set of tools that include hardware and software, data base management, land-related data, the logical structuring of topology into points, lines, and polygons, and the spatial analysis capabilities that make geographic information systems unique in the computer graphics industry. The geographic information system is compared to computer-aided design and drafting (CADD), computer-assisted mapping (CAM), and automated mapping and facilities management (AM/FM) systems.

Chapter 3 discusses various applications of urban geographic information systems in service delivery, management, and policy activities today. Categorizing by administrative level, and citing examples of case studies and reproductions of actual uses, this chapter provides a valuable source of information for the doubtful—because these applications have been proven to work. This chapter is supplemented by the Applications Digest in the back of the book, which gives a more comprehensive, yet less descriptive, list of applications.

Chapter 4 helps the student of GIS understand the topological data of a local government and why these data are important in a GIS. Points, lines, and polygons are all that the computer has to work with, because it does not have the cognitive ability to "see" relationships on a visual image as do humans. This chapter introduces "GIS County," a comprehensive exercise in developing an urban GIS that is continued through the remainder of the chapters and is helpful in understanding the concepts they present.

Chapter 5 provides the student with the basic concepts essential to using a geographic information system in an urban environment. The principles of the geographic base file (GBF), taken from urban topological data structures, help the student understand the basic geographic information needed in a GIS, including descriptions of DIME and TIGER Files available from the U.S. Bureau of the Census. The GIS County exercise gives the student experience in creating and using a GBF.

Chapter 6 describes the land records information that drives the GIS. Geodetic reference of parcel-based maps and the land-related information generally computerized by many offices in local government today are the essential elements for using a GIS to solve problems, answer questions, and produce accurate maps. GIS County allows the student to process and analyze parcel-based land-related data.

Chapter 7 presents a model that should be emulated by all urban geographic information systems projects. This model includes the processes of identifying the geographic information needs in an organization, justifying the project to the financial decision makers, conducting a pilot project to test and evaluate the use of the system, converting existing records to digital form, and managing the implementation of the system.

This book is modeled after the syllabus of an introductory GIS course I taught in the fall of 1988 at the University of Wisconsin-Milwaukee. The students in this course came from diverse graduate-level studies— Urban Planning, Geography, and Public Administration—and learned, I think, that a geographic information system used in an urban setting can be many things to many different people involved in the running of government. As the material in Chapter 3 attests, I discovered this at the City of Milwaukee during the past fifteen years of managing a GIS for engineers, planners, administrators, managers, and elected officials. Establishing one of the nation's first urban geographic information systems in 1975 (before it was even called a GIS), gave me more challenges than I sometimes want to think about in applying this technology to urban issues.

My education surely prepared me for these challenges. My undergraduate industrial engineering studies at Northwestern University and graduate engineering management program with public administration specialty at the University of Dayton gave me a foundation for improving systematic processes in public agencies. A few years of summer employment during college as a surveyor's rodman provided me with a basic understanding of how maps are created and the value of accurate survey records. Now, after a year's employment in the city's planning department and currently working in the data processing department, I have seen both the user side and the provider side of information systems. My involvement with the Urban and Regional Information Systems Association (President, 1984–85) has committed me to improving the use of urban information and to educating the vast group of dedicated public officials who are faced with data needs daily for addressing public issues.

We need more GIS-educated professionals in local government. Whether they run the government or whether they run the GIS project, they need to know the full capabilities of geographic information systems. Read, for a moment, what one former student, also a public administrator, wrote, in the summary of his term paper, about the need for GIS education:

> One of the most interesting things I've learned from doing this project and from taking the class this semester is the unlimited potential GIS has for not only the Bureau of Sanitation, but for most City departments. Life could have been so much easier for (garbage) route devleopment, (garbage) cart implementation, and other bureau projects had I been more familiar with GIS capabilities. It's caused me to wonder how many other

"bureaucrats" there are in the City system, that for lack of knowledge, leave a resource like GIS untapped. If I leave you with one recommendation, it's to continue to get out there and sell your services. I think you've begun to see the acceptability among the elected officials. Now all the bureaucracy needs to be educated.

To "the bureaucracy" and those about to become it, I urge you to study this book.

Milwaukee   W.E.H.
November 1990

Acknowledgments

This book could not have been written if not for a special group of individuals—a group of dedicated professionals from diverse backgrounds who, between 1974 and 1987, were a team: The policy development information system (PDIS) team. For many years the members of the PDIS team did not know we were building an urban geographic information system; the term *GIS* was not widely acknowledged back in the early days of development at the City of Milwaukee. Our charge was to make computerized property information more available and easy to access by the workers, the managers, and the policymakers at the City. Our dedication was to those city employees who needed information to improve their jobs—the geographic information system came later. From the work of these people (some have moved on either in this profession or in some other productive endeavor), a GIS was formed and was proven to be a valuable resource in an urban environment. Although they may not know it, the following PDIS team members and others who were instrumental in supporting it, helped me write this book:

Rich Allen	Frank Bayer	Bill Drew
Eileen Francis	Randy Gschwind	Randy Guyer
Bernie James	Greg James	Bob Juhay
John Kay	Reinhold Knopfelmacher	Susan Kronberger
Mike Marley	Mark Moss	Paul Mueller
John O'Donnell	Bob Polikowski	Aqil Rajput
Terry Requarth	Bill Shaw	Randy Sincoular
Sharon Struble	John Sullivan	David Turnpaugh
Rose Williams	Kathy Bertrand	

    My special appreciation goes to Dr. G. William Page, acting dean of
the Graduate School of Architecture and Urban Planning at the University of Wisconsin–Milwaukee, who not only was responsible for establishing a series of GIS courses at the university but who also one
day said to me: "You ought to write a textbook." (Dr. Page is now
dean of the College of Urban and Public Affairs at Florida Atlantic University.) Others who provided valuable and critical encouragement in
the writing of the manuscript for this book include Dr. Kurt Bauer and
Thomas D. Patterson at the Southeastern Wisconsin Regional Planning
Commission (SEWRPC); Dr. Robert T. Aageenbrug at the Geography
Department of the University of South Florida; D. David Moyer at the
National Geodetic Survey; and Dr. Joseph Ferreira at the Computer
Resource Lab at the Massachusetts Institute of Technology.
    Finally, I am forever indebted to two fine institutions that provided
me with both the resources and the opportunities over the years to
develop material eventually used in this book: the City of Milwaukee
(and its Common Council, previous mayor the Honorable Henry A.
Maier, and current mayor the Honorable John O. Norquist) and the
Urban and Regional Information Systems Association, with Tom Palmerlee as executive director.

Contents

3. **Applications of Urban Geographic Information Systems** 64

Map Updating, 65
  Plat Maps, 65
  Quarter-Section Maps, 67
  Land Use Maps, 69
Zoning, 70
  Recording Zoning Changes, 71
  Enforcing Zoning Restrictions, 77
Reapportionment, 79
Building Inspection Workload Balancing, 83
Solid Waste Collection Routes, 93
Public Policy—Housing, 95
  The "Dirty Dozen" Housing Code Violators, 95
  Boarded-Up Properties, 101
  Tax Delinquent Properties, 105
Liquor License Policy, 112
Health Policy on Lead Poisoning, 117
Library Facilities Planning, 120

4. **Topological Data Structures** 127

Graphical Representation of Cartographic Data, 127
Linking Nongraphics Data to a Cartographic Data Base, 132
  Display of Nongraphic Attribute Values, 134
  Display of Selected Nongraphic Attribute Values, 135
  Changing Map Features Based on Nongraphic Attributes, 136
Adding Topological Data Structures to a Cartographic Data Base, 137
  Point Tables, 138
  Line Tables, 139
  Area Tables, 140
  Examples of Using Topological Data Structures, 140
GIS County Exercise 1: Topological Data Structures, 142

5. **Geographic Base Files** 147

Geocoding and Geoprocessing, 147
Elements of a Geographic Base File, 157
  Street Segments, 157
  Geographic Areas, 159
Types of Geographic Base Files, 163
  DIME Files, 163
  TIGER Files, 166
  TIGER/Line Files, 171
GIS County Exercise 2: Geographic Base Files, 180

An Introduction to Urban
Geographic Information Systems

1

Information in the Organization

> We have for the first time an economy based on a key resource
> that is not only renewable, but self-generating. Running out of it
> is not a problem, but drowning in it is.
>
> NAISBITT, p. 24

The observation by John Naisbitt that information is a resource to our economy just as people, money, and equipment are resources for manufacturing goods, has led many large organizations to manage their information better. That is, many organizations have begun to realize that they have invested an enormous amount of money in the data that are stored in their computers and have implemented new standards, policies, and procedures to help control the costs associated with computerized information systems.

When computers were first introduced to these organizations in the 1960s, they allowed information to flow more efficiently through the organization. The processing of time cards into paychecks, for example, was faster on the computer and required fewer clerks to complete the process. This improved efficiency because the processing of the payroll could be done more rapidly and with fewer people. When managers later began to realize that the information collected on the time cards to process payroll checks could also be used to measure the amount of time spent on specific tasks, they began to ask for computerized printouts of work-hour totals by task. These statistics allowed them to evaluate the tasks of their organizations and make improvements in the way they conducted their business. Thus, computerized payroll information used to prepare paychecks in order to improve the *efficiency* of the payroll process, began to be used by managers also to improve the *effectiveness* of the organization as long as the computer could produce and process the information needed by the managers.

By improving both efficiency and effectiveness, computer processing technology added value to the information that was computerized. This value was associated not only with the efficiency of the workers, but also with the effectiveness of their efforts and of the efforts of the managers and policy-makers of the organization.

The Value of Information in the Organization

Twenty to thirty years after the computer revolution automated many information-processing tasks to improve the efficiency of the information workers in an organization, the ability to enhance worker effectiveness began to drive computer usage rather than the ability to make them more efficient (Strassmann, p. 117). Inspecting more garages this year than last is more efficient, but not very effective if the *houses* that were not inspected (because so many garages were inspected) had serious safety defects. Strassmann asks: "What good is an efficient car (one that gets 100 miles per gallon) if it has no place to go?" How effective is the use of that car? Improving not only efficiency, but also the effectiveness of the organization is why computerized information is valuable.

Another reason why computerized information is valuable is because it can be shared with other functions in an organization if it resides on a centralized computer. In exchange for manually researching another department's file cabinets, a computer terminal connected to the computer with access to the other department's data can be a valuable asset. Why should a building be demolished because the Building Inspection Department has issued a condemnation order when it has just been declared an historic monument by the Planning Department? If the information that each department has on the same building is shared in a timely fashion, these mistakes can be avoided, and the whole organization can be more effective. This is evidence that the value of information increases the more it is shared and disseminated. On the contrary, information that is not used is useless (Strassmann, p. 117). Take, for example, the city that directed its police officers to record all the information they received from field interrogations (questioning someone in suspicious circumstances). The officers completed forms containing the name of the person, the time and date, the location, and a description. These forms were then submitted to the sergeant at the end of the shift and sent to Headquarters to be stored in a series of boxes—thousands of forms each day. Talk about "drowning in information"? This information was never used by detectives investigating crimes because looking through all those forms was physically impossible. The information recorded by the police officers on field interrogations was useless and therefore had little, if any, value.

Public organizations such as municipalities and counties experience a rapidly changing political environment that can cause radical differences in the way public employees work. The computerized information systems designed to make them more efficient must be flexible enough to allow for changes in their work and in the information that they use and provide to others. This rapidly changing environment of public service also requires information systems that are flexible enough to provide critical information to the policy-makers as their needs change almost daily. Information systems designed to automate procedures to improve efficiency, however, have rigidly defined data structures and programs that cannot deliver the full potential information has to offer because they are not flexible.

Traditional Information Systems Design

Information systems designed to automate existing manual procedures are called *transaction-based* systems. They take a transaction (an input record) and process it through a series of programs that manipulate the data on the record, retrieve additional data from tables and files stored in the system, and then write new information to other files. Finally, after the data have been manipulated accurately and completely, a product is produced by printing necessary data on forms or tabular listings. The transaction has been input to the system, processed by programs, recorded on files, and then output in a different form, as shown in Figure 1.1.

Think of a large organization's payroll system. Hundreds or thousands of time cards are completed by the employees every 2 weeks or so. These transactions, the time cards, are converted into computer-readable form and input to the payroll system (see Figure 1.2). Programs *read* each transaction, *access* tables and files for additional information (such as rate of pay by type of employee or rate of Social Security deduction, etc.), and then *perform calculations* on the input data based upon this additional information. The transformed data (from hours to dollars) are then *recorded* on a new file for historical purposes. The final step is to *output* these transformed data onto paychecks.

FIGURE 1.1. A transaction-based information system. A computer program reads data from an input transaction, retrieves data from a file, performs calculations, writes new data back to the file, and then prints an output document.

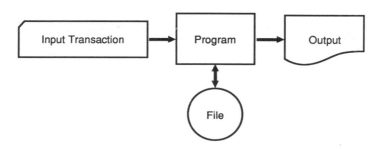

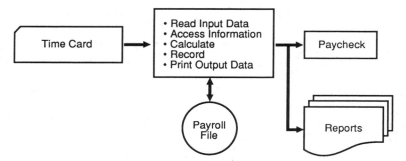

FIGURE 1.2. A transaction-based payroll system. Computer programs read payroll information from a time card, find the rate of pay for the employee from a payroll file, calculate a paycheck amount and deductions, record the calculations onto the payroll file, and then print checks and management reports from the results.

Many city or county tax assessment systems are transaction-based because they take data input from the results of an assessor's inspection of a property, access tables and files to obtain figures that can translate property characteristics into assessment values, perform the calculations, write the new assessment values to a tax roll file, and then print an assessment notice that is sent to the property owner (see Figure 1.3).

Most organizations have transaction-based information systems because they are efficient in processing information (not as many clerks are needed to process paychecks, and property assessments are computed more rapidly than by hand). Other reasons why transaction-based information systems are so popular are that:

FIGURE 1.3. A transaction-based tax assessment system. The property characteristics are input transactions that are used by computer programs to retrieve data about each property and valuation rates from a tax roll file. The programs then calculate a new assessment value for each property and record this new value back onto the tax roll file. Finally, a property assessment notice is printed and sent to the owner of the property.

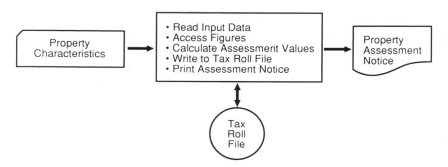

- They are easy to cost-justify because their use can be shown to reduce the cost of government.
- They are relatively quick to develop because they are designed to serve one purpose.
- They are easy to use because they are relatively simple in design.

Now that users have become more sophisticated in their understanding of computers, and now that computer technology has begun to offer more capabilities, however, these transaction-based information systems are becoming more of a problem than a solution to the information needs of government. Managers are realizing that data that have been computerized to improve one function can also be useful in another function and therefore, have begun to ask the data-processing center to make modifications to their systems. Since the conceptual design of the original systems was based upon single functions, their structure was rigid and difficult to change to accommodate new functional needs (see Figure 1.4). System modifications were complex and time-consuming as "patches" were installed to "fix" complex situations. As more and more changes were requested, the data processing center began to spend more time changing existing systems than in developing new ones (Martin and McClure, p. 25).

Figure 1.4 portrays a transaction-based tax assessment system (such as that shown in Figure 1.3) expanded to provide additional capabilities beyond the calculation and creation of property tax assessments. Four additional programs and two data files have been created to allow additional update, retrieval, and summarization of data for purposes other than producing tax assessment values. If, after the modified system has been in operation for some time, the Owner ZIP field is expanded from the traditional five-digit ZIP Code to the new nine-digit ZIP+4 Code, all programs and two files must be changed to accomodate the larger-size field. Program 1 must be changed to allow the entry of a larger number, and Program 2 must be changed to allow retrieval of the larger number. File 2 must be expanded to provide room for the larger field. Since Program 3 uses File 2, it also must be changed because the size of the file has changed. This, in turn, causes a change to File 3 because it contains the value from the Owner ZIP field of File 2. Because File 3 has been changed, Programs 4 and 5 must be modified to allow for the larger size of the field. Even though Programs 3, 4, and 5 do not process the values of the Owner ZIP field, they must be modified because the size of File 2 was changed. The only part of this system that does not require modification when ZIP Codes expand to nine digits is File 1, which does not contain the Owner ZIP field.

As government services become more complex in the midst of cost-reducing property tax relief efforts, individual departments have come to depend upon each other more and work together on certain services and programs. This has caused managers to share computerized data

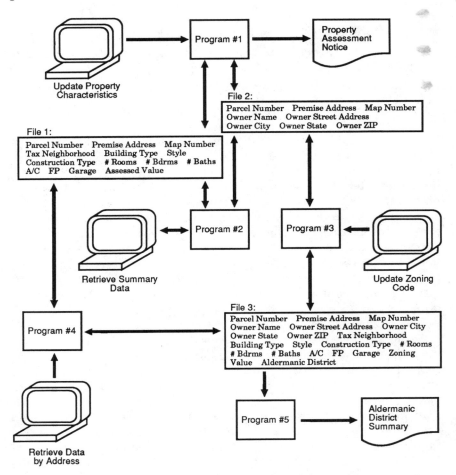

FIGURE 1.4. A transaction-based tax assessment system modified for additional uses. Here Program #1 reads, calculates, records, and prints property assessments as does the program in Figure 1.3; however, four additional programs and two data files have been added to the system so that the data can be used for other purposes (such as producing summary statistics, updating zoning codes, retrieving data by address, and printing tax assessment summaries by aldermanic district).

when the functions they perform process information that can be useful across departmental boundaries. They have begun to realize that it is the *data* that are valuable to share in these systems and that the original purpose of automating procedures was not the only reason for using computers. The data-processing center again became overwhelmed with requests from users to transform their systems from transaction-based systems to data-based systems. The original design concepts, however, were not flexible enough to adjust to this major change in philosophy. Almost insurmountable problems were encountered: coordinating data update cycles; keeping track of the many files, tables, and programs needed to integrate data among systems; translating different data codification schemes between departments; and providing easy access by the user to the data when they were not sure exactly what was needed.

The Data-Base Solution

In the 1970s and 1980s, the data-processing profession addressed these problems by developing *data-base management systems* (DBMS). Data-base management systems make the data independent from the programs, applications, and systems used by the many different functions in an organization (Martin and McClure, p. 114). Using a data base, or set of data bases, which are managed separately from the programs that access them, DBMSs allow changes to be made to either the data or the programs without necessarily causing changes to be made to both (see Figure 1.5). Thus, when the five-digit ZIP Code changed to the nine-digit ZIP+4 Code, only one field on the data base required a change—not a whole file. Only those programs that used ZIP Code had to be changed—not every program in the system. This reduced program maintenance significantly.

DBMSs also enhance data sharing among different functions in the organization. Since the design of the information system focuses on the data in a DBMS environment, the information needs related to the subject matter of the data base are considered on an organization-wide scale rather than on one individual office. This reduces the many files and subfiles inherent in complex transaction-based systems to only one or a few data bases. This organization-wide perspective on data-base design promotes the logical linking of data from diverse functions when necessary. It also allows better control over update cycles because fewer files are involved and because diverse functions depend more on each other and communicate better in a cooperative environment. In other words, different offices and departments are more or less forced to cooperate on data management when their information systems are integrated under a DBMS environment.

Why are all information systems not designed with data-base concepts and DBMSs? Because they are expensive, are time-consuming to develop, and require a high level of sophistication to use. The new DBMS technology has taken many years to develop at considerable expense on the part of the companies offering these systems. They must recover those investment dollars by setting prices for them in the hundreds of thousands of dollars for mainframe computers (and that does not include the cost of the computer!). The DBMS technology continues to evolve. One of the main criticisms with current DBMSs is that they are not "user-friendly." So much effort has been expended on the

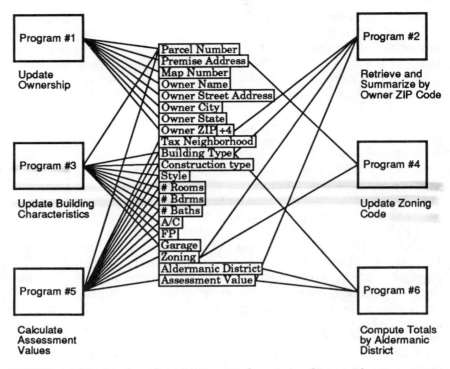

FIGURE 1.5. A data-based property records system. Since each program accesses only those elements it needs, a change in the size of an element affects only those programs that use it. Here, the element owner ZIP is expanded to accommodate the nine-digit ZIP + 4 Code, and only two programs, Programs #1 and #2, require modification. The remaining four programs do not use the Owner ZIP field; so they do not have to be changed.

technical perfection of logical and physical data structuring (see Chapter 2 on hierarchical versus relational data-base structures) and programming functions (including data security, transaction logging, disaster recovery, etc.) that good user interfaces have not yet been adequately addressed. While current research and development of *fourth-generation languages* (4GL) has improved somewhat the ad hoc access to data bases by users, there is still much work to be done on making DBMSs more friendly to the casual user. As a result, managers have begun to train specific individuals within their own departments to maintain expertise in using these systems.

The development cycle needed to implement data-based information systems is much longer than the development cycle for transaction-based systems. This is not because of their technical complexities so much as the complexities associated with the integration of data among diverse operating functions of the organization. A corporate data base (one that is used by all or most operating functions of the organization) cannot be built and successfully used until all possible users have been interviewed and involved in its design. Coding standards and data definitions that differ from office to office must be modified and understood by all users. Data and operating standards must be developed and understood by all users. Since users are more involved in accessing the data, comprehensive and understandable documentation must be written. All of the work in interviewing potential users, developing standards, and writing understandable user documentation takes time and therefore lengthens the development cycle for these systems.

One of the biggest problems in implementing data-based information systems is related to the high cost of developing them: They are difficult to cost-justify. The traditional methods used to cost-justify transaction-based information systems have been to reduce staffing levels (and therefore operating costs) because they introduce efficiencies into the organization. Data-based information systems are not only more expensive to develop than transaction-based information systems, but because they are a benefit to many different functions of an organization, their use cannot always be related directly to cost savings. This may sound counterintuitive at first, since more users deriving benefits from a system should mean more efficiencies, but those efficiencies do not always result in reduced staffing levels. A ten percent reduction in work in a user's job does not translate into a ten percent reduction in cost. Ten users from ten different departments saving ten percent of their time each does not add up to one hundred percent of one worker's job. These people have other work to do. The fact that data-based information systems make many workers (managers and policy-makers included) more effective in their work is much more difficult to translate into cost savings than the fact that transaction-based information systems introduce efficiencies into the organization. As long as decision-makers follow traditional thinking that the cost of information systems should be offset by dollar savings in efficiencies, data-based information systems will be difficult, if not impossible, to justify. (More on this in Chapter 7.)

The problems with data-based information systems notwithstanding, they are popular, they are successful, they are demanded by users, and they will flourish as the cost of the technology is reduced, as data-processing professionals become more efficient in developing them, and as user sophistication and user-friendliness converge. The effectiveness of these systems will play a larger role in the "go/no go" decision process as decision-makers begin to understand the value of information and the cost to the organization of not having it when it is needed. We know this will happen because new methodologies are being developed in the data-processing profession to help designers of information systems link data-base development more directly to the strategic plans of the organization. *Data-modeling* techniques are now being used by many organizations to develop systems that can be shown to be strategic necessities for the organization to meet their goals. This will weed out the costly systems that appear to be beneficial, but are not critical to the realization of the goals of the organization.

Data-modeling concepts concentrate on the *information needs* of all levels in the organization rather than on the *data-processing needs* of individual functions that are subject to change repeatedly as the political and legislative environment of government changes rapidly.

The Urban Information Pyramid

The organizational functions of governments are no different from those in private industry: A service or product delivered to customers through the management of its resources and under policy guidelines established by the organization leaders. Thus, the major functions of an organization, public or private, are:

Operations—producing or delivering the product or service;

Management—controlling the organization's resources needed to run its operations;

Policy—establishing the long-term, overall direction of the organization.

Government operations include delivering water to city residents and taking away their wastewater, providing food stamps to those who qualify and collecting taxes from those who own property and many, many more service-related activities. Public managers use money, people, and time, within the constraints set by the policy-makers, to orchestrate the activities that perform those government operations. The policy-makers, accountable directly to the public, provide the resources and constraints to the managers that reflect the will of the people. All three levels of government, operations, management, and policy, perform the functions that the organization needs to survive.

Figure 1.6 provides a graphic representation of these functions. They can be viewed as levels within a pyramid with operations at the base, management in the middle, and policy at the top.

The base of the pyramid, the operations level, is where government policies and management expertise produce actions that affect the public. For example, city residents are provided with garbage carts by the city after the policy-makers have decided to begin a new program for collecting garbage and after the managers have obtained the carts, trained the collectors on their use, and established new collection routes. The base is the largest level of the pyramid because there are more people (garbage collectors, tax assessors, building inspectors, police, etc.) at this level than at others.

The middle of the pyramid is shown as the management level of the organization because this is where policies (budgets, programs, ordi-

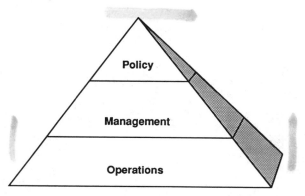

FIGURE 1.6. The government business pryamid. Policy level functions address organization-wide issues over multiyear time periods. Management level functions translate policies into actions through the efficient and effective use of the organization resources. Operations-level functions deliver services to the public and also support these service-delivery activities.

nances) set at a higher level are formulated into action plans. Resources used in the operations level are managed at this level.

The top level is where policy is established—usually by elected officials. Budgets are approved; laws and ordinances are enacted; new programs and other policies are established. This is the smallest level because there are fewer people involved in this function.

The pyramid in Figure 1.6 is shown in three dimensions because behind each level of the organization is information that supports that function. Through *horizontal* and *vertical data integration,* data used to support the operations level is combined with other information (horizontal data integration) and summarized (vertical integration) as it flows up the organization. Information gathered at the operations level, then, is eventually used by policy-makers to initiate plans and programs that, in turn, are sent down to the managers where they are formulated into actions for the operations level to complete (see Figure 1.7). Effective policies and programs are dependent on accurate, comprehensive, and timely information.

Successful information systems that support these levels of local government are those that are based upon the data needs of its operations—its service-delivery functions. The reason for this is simple: Good decisions require good information. The operations level of government (the tax assessors, building inspectors, meter readers, nurses, etc.) consists of the people who work with and depend on data daily to perform their tasks. If their information is bad—inaccurate or out of date—then their jobs are more difficult to perform. Since they need the data to perform their job, they will ensure that they are accurate and current. If, however, they are required to collect data for some other function not related to their job, then the incentive to ensure accuracy

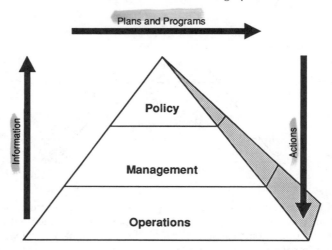

FIGURE 1.7. Information flows up the government business pyramid, where policy-makers approve plans and programs to cause management actions that affect the citizens.

and currency is not as great as if they depend on the data themselves.

Data are created by the operations level of government: Tax assessments are made; payments are received; buildings are inspected; meters are read; etc. Many operations of government record information as part of their function. Not all of these data, however, are needed at the management level. Managers generally need summary information of the operational data. Instead of house-by-house or parcel-by-parcel data, aggregate information (by employee, by district, by program, etc.) is needed by managers to ensure productivity, workload balance, and proper schedules. Even fewer data are needed for policy purposes. At the policy level, the selection of information is difficult because it must come from many different operations. The questions that are asked shift from "how many inspectors are needed?" to "should the city demolish and rebuild or rehabilitate deteriorated houses?" Less detailed data are needed, but more integration, aggregation, and flexibility are needed to meet the ever-changing information needs at the policy level.

Operations Example

An example of the information needs at the operations level of government can be found in the building-inspection process (see Figure 1.8). Here, specific tasks (inspecting houses and businesses) are performed daily. The building inspectors, usually assigned to a certain geographic area (a building-inspection district), record the addresses of the buildings they inspect and any building code violations found. Later, notices are sent to the owners, ordering them to correct the deficiencies.

Bureau or Divisional Concerns Which Result In Specific Tasks:

- Inspections
  - What are the violations?
- Reinspections
  - Which homes require reinspections?
- Enforcement
  - How long have violations been outstanding?

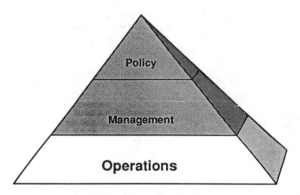

FIGURE 1.8. An example of operations-level activities in building-inspection-related functions. These activities include specific tasks related to inspecting properties. Information needs are specific: address, date, violation, etc.

In addition, the inspectors must reinspect these properties after a certain time period to ensure that the corrections have been made. They must know where to go, what to reinspect, and how long each violation has been unabated in this reinspection process. Their task is to enforce building codes by inspecting, recording, notifying, and reinspecting. Their information needs are very specific: addresses, violations, dates, etc.

Management Example

The managers of the building inspectors use the information collected during the inspection process to direct departmental resources in fulfilling their mission (see Figure 1.9). The department head must prepare an annual budget request based upon how many inspectors are needed to enforce building codes. The supervisors of the inspectors must evaluate the performance of each of their subordinates (number of inspections, time spent on inspections, number of reinspections, etc.). All department managers assist other officials in developing special programs for intensive treatment or financial assistance in areas of the city with high incidents of violations or the most serious types of violations. They constantly review workload statistics to ensure workload balance and make adjustments where necessary. The information they

Departmental Concerns Which Affect the Ability of the Department to Fulfill Its Mission:

- Budgeting
  - How many inspectors are required to inspect X number of homes?
- Performance Evaluation
  - Are all violations being re-inspected?
  - How many re-inspections are required?
  - How long is compliance time?
- Housing Program Development
  - Where are the high incidents of violations?
  - What percentage have serious defects?
  - What would it cost to repair them?
- Workoad Balancing
  - How large should an inspection district be?
  - Do all inspectors have the same amount of work?

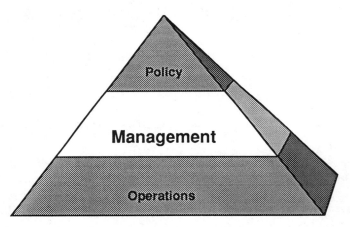

FIGURE 1.9. An example of management-level activities in building-inspection-related functions. These activities involve managing people and money on a department-wide basis. Information needs are statistical: totals by inspector name, inspection district, type of violation, etc.

use is summarized by inspector name, inspection district, time period, type of violation, and other categorizations.

Policy Example

The policy-makers and other policy-related staff address city-wide issues affecting all departments over multiyear time periods (see Figure 1.10). The annual budget process requires them to compare the services of different departments and make decisions on service levels and programs given a limited amount of resources to allocate. Issues addressed may include: crime prevention versus housing rehabilitation; tree planting versus financial assistance to homeowners; or health inspections versus housing inspections. Federal funds must be allocated to programs proposed by various departments and the public. Ordinances are proposed that would change zoning or housing enforcement procedures in the department. The analyses necessary to make these types of decisions require less detailed data but more flexibility in the use of the data ("what if" types of questions). Data for managers and policy-makers must be comprehensive and integrated with data from other functional responsibilities in the government.

Figures 1.8, 1.9, and 1.10 do not represent three different information systems for the Building Inspection Department. They represent three different functional uses of the same information system—a housing code inspection system (see Figure 1.11). The design of the system shown in Figure 1.11 is such that it contains standards and flexibility so that it can be used at all three levels. Its features are *accuracy* at the detailed level (and hence, all other summary levels), *currency* because the inspectors depend on it to do their job, *efficiency* because there is less data redundancy than having three separate systems, and *durability* because it is needed by so many people. Figure 1.12 portrays

FIGURE 1.10. An example of policy-level activities related to building-inspection functions. These activities involve the future of the government as well as the citizens: Budgets affect taxes; new programs affect business and living conditions; etc. Information needs are comprehensive and unstructured, requiring integration among many different functions and flexibility for use in analyses.

Citywide Concerns Which Direct City Services Over Multi-year Periods:

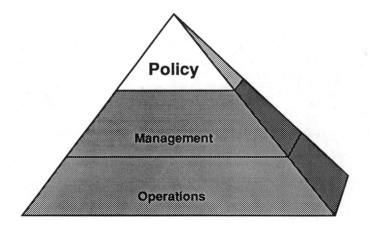

- City Budget Process
- Community Development Program Approval
- Housing Code Ordinance Changes

Policy

Management

Operations

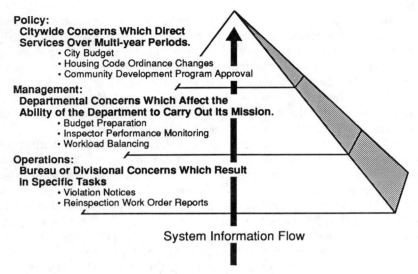

FIGURE 1.11. A housing code violation system designed to meet the information needs of the operations, management, and policy levels of government.

FIGURE 1.12. A property records system designed to meet the information needs of the operations, management, and policy levels of government.

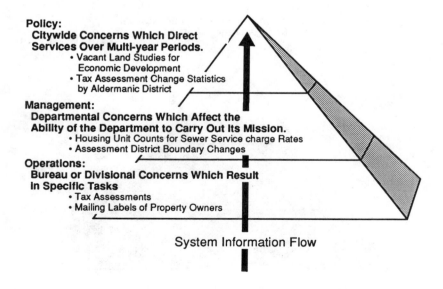

another such system: a master property file system based on tax records.

The goal, then, in developing information systems is to design ones needed at the operations level of government, but also with the necessary standards and flexibility to allow the data to be integrated with other data in different functions of the organization and aggregated in summary form for assimilation at the management and policy levels.

Ineffective Systems

Ineffective systems are those designed only for operational activities or only for management or policy purposes. The former cannot be summarized or integrated with other data systems for use at higher levels of the organization. The latter require data to be created outside the normal operating activities and do not have the support they need when workload demands in other tasks increase or when budgets get trimmed.

Figure 1.13 represents an information system designed only for an operational activity. Such a system is single-purpose in nature and not easily modified for other uses. Without data standards it cannot produce summary information nor be integrated with other information systems. Because it is designed for only one purpose, it is not flexible enough to be used for other purposes. The result is that information cannot flow easily up the pyramid across the operations boundary for use by management and policy personnel.

Take, for example, the East Coast city that implemented an automated mapping system to improve the productivity of drafting personnel in the Public Works Department. The system was installed, and all of the map sheets containing public right-of-way and platted lot

FIGURE 1.13. A system designed for an operations level without consideration of management- and policy-level needs. Lack of flexibility and data standards prevents the use of the system for purposes other than a single use. Information cannot be used by managers and policy-makers.

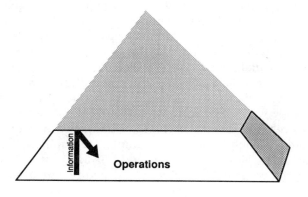

boundaries were converted to digital form on the computer. The system produced the productivity improvements originally anticipated for the daily operations associated with map updating by the drafters, but, when the Planning Department needed a map of the entire city for a planning study, it could not be produced on the system. Since the digital map sheets were stored in separate files with no coordinates to locate them within the entire geographic area of the jurisdiction, they could not be combined to form a map of the city. While the daily map sheet updating functions of the Public Works Department became more efficient by the use of automated mapping technology, other functions of other departments requiring maps of large geographic areas (census tracts, inspection districts, aldermanic districts, etc.) for planning, management, and policy activities could not make use of the new technology.

Another city decided to automate the process of writing letters to property owners after inspectors found building code violations in their buildings. An optical scanning device (much the same as those used to score tests) was installed to read specially designed forms marked by the inspectors during the inspection process. Each evening, all of the forms that had been completed during the day were fed into the scanner, which then read the marks, translated them into specific building code violations, and printed a letter notifying the property owner of each code violation found and when it had to be corrected. The process made the department more efficient because it allowed the inspectors to double the number of inspections they could make, and, at the same time, it reduced the size of the clerical staff from six typists to two. However, since the system was designed only to convert the marks and print the letters, the information from the inspections was not stored in the computer for other uses such as: keeping track of buildings that needed reinspection after the allotted time for correction was up; keeping track of buildings and property owners that had recurring violations; and analyzing the types of violations recorded for determining special housing programs that could be initiated for improving the city's housing quality.

Conversely, Figure 1.14 represents an ineffective "management information system" or "decision support system." It is not based upon data collected as part of an operational activity, and therefore does not have the same type of support as the examples described earlier. In such systems, management typically dictates that certain data are to be captured by operational personnel and entered into a management information system. Since the data are not needed by the operational personnel, they will not be accurate and current. The system will eventually die as other workloads increase or budgets are trimmed.

One large-city comptroller, for example, determined that a "fixed-asset management system," nicknamed "FAMS," was needed to improve the management of all city-owned equipment used in the var-

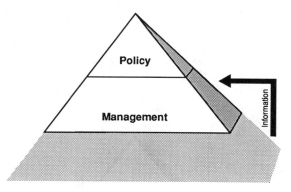

FIGURE 1.14. A system designed for management and policy needs without consideration of operations-level needs. Because the data are collected only for managers and policy-makers and are not needed in an operational, on-going activity, accuracy and currency will suffer.

ious departments of the city. The FAMS system was to track every equipment item purchased by the city with a value of more than $100 (desks, chairs, bookcases, computers, automobiles, trucks, etc.). The information collected would be used to reduce equipment purchases by being able to reallocate used equipment no longer needed by one department; determine the total value of all equipment for making better decisions on insurance matters; and evaluate the quality and durability of purchased equipment for improving future purchases. The FAMS system was billed as a true management information system because it was to be built for the purpose of managing information. Coding forms were sent to each department for recording the necessary information about each piece of equipment with a value of more than $100. Clerical personnel were assigned the task of collecting the information and coding the forms, but because the clerks had to do this in addition to their other responsibilities, they often did not have enough time to collect the information. The coding sheets were never completed because their other duties were more important, and they did not need the information to perform them. The system was a failure.

Some cities or departments within cities have implemented a geographic information system to solve a particular problem or address a specific issue related to public policy and have collected pertinent information, digitized maps from various sources, and analyzed the data for the project or program. Later, when the problem or issue was no longer important, the system could not be used for other purposes because the data were not kept current and the analysis programs were not generalized for other uses. Without a solid foundation for data maintenance on a regular basis as part of the operations-level functions of the organization, a geographic information system will not be an effective tool for use by managers and policy-makers.

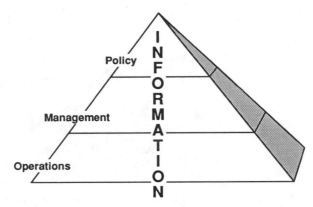

FIGURE 1.15. Successful information systems are designed to improve the operations level of government and also support the information needs of policy and management levels.

Data processors should strive to develop systems that not only improve the daily operations of government, but can also be used by managers and policy-makers to improve decision making, planning, and policy analysis. To accomplish this, they must understand the needs of all levels of the government and design systems with strict data standards and flexible use. Figure 1.15 presents a picture of how data needed at the three levels can support the functions of each when information systems are designed from the perspective of the entire organization rather than one particular function. These systems have solid foundations, are built to last, and have multiple uses. They are the urban information pyramid.

Geographic Information in the Urban Environment

Outside of financial accounting and purchasing information systems, it is difficult to think of urban information systems that do *not* process data that are location-related. During the years 1985–87, the Wisconsin Land Records Committee, a 32-member committee with an additional 71-member subcommittee structure, was charged by then-governor Tony Earl to study local, county, state, and private land-related information needs in Wisconsin. The committee's final report (Wisconsin Land Records Committee, p. 1) found that:

> . . . of all information collected and maintained by all levels of government, the percentage that is land-related is so large that it is difficult to imagine a data set whose value would not be enhanced by a geographic or locational reference.

A 1986 brochure (Municipality of Burnaby) published by the Municipality of Burnaby, British Columbia, reported the results of a needs analysis for an urban geographic information system (GIS) in that municipality: eighty to ninety percent of all the information collected and used was related to geography.

The geographic nature of data processed in today's cities and counties provides almost unlimited opportunities for improving service delivery, management, and policy-making in these governments. This is because geographic data provide natural linkages that allow data to be integrated horizontally across the organization (addresses from tax records can be related to addresses from building-inspection records; census tract statistics from the census data can be related to census tract statistics from crime records; etc.) and allow data to be aggregated vertically up the organization levels (crime data, tax assessment data, water-main break data, and other data collected by address can be summarized by census tract, aldermanic district, etc.). As Chapter 3 will show, GIS technology applied to our urban problems allows new ways to use the enormous amount of data already in computerized form.

Geographic information systems technology is causing data-processing professionals to rethink their solutions to the information problems of today's public officials. They hear GIS professionals talk about "geographic data bases"—not "automated systems." They are asked by users to automate information—not procedures. Mercer and Koester (p. 101), in their discussion of management information systems for public agencies, call for a "comprehensive viewpoint" towards information managment—one that spans organizational levels with multifunctional applications. They specify five recommendations for policy-makers, the first four relating to the use of data-base management, standards, and distributed processing. Their final recommendation for policy-makers is: "Public sector organizations should eventually utilize a CODASYL data-base management system for structuring and manipulating a geographic base file." The reason they give for this recommendation is that "by utilizing a generalized data-based management system, geographic data can be easily linked to other data used by the organization."* (Note: CODASYL standards for data-base management systems were developed by a task group of the Conference on Data Systems Languages.)

Traditionalists see applications of this new technology as a means to introduce efficiencies in government and use the features of electronic drafting as justification for investing in GIS resources. But automating the process of map making is only the beginning when it comes to adding value to the information processed by an urban geographic information system for needs at all levels of the organization. This observation is made by Jack Dangermond, president of Environmental Systems Research Institute (ESRI), who says in the "PC/ARC/INFO Starter

*Reprinted by permission of publisher from *Public Management Systems*, by Edwin Koester and James Mercer © 1978 AMACOM, a division of American Management Association, New York. All rights reserved.

Kit" video (Part I) that there are three goals that geographic information systems strive to meet:

1. Increased productivity in utilizing maps and geographic information;
2. Improved geographic data management; and
3. Better strategic ways to use geographic data to support decision making.

Exercises

1. Why are computers used to process information?

2. Explain the difference between a transaction-based information system and a data-based information system. What are the advantages and disadvantages of each?

3. Data integration is the combining of different data bases to satisfy user information needs. Explain how data can be integrated horizontally and vertically in an organization such as a municipality.

4. Describe how crime data can be used at the operations, management, and policy levels of local government. (Hint: Information is collected at various stages of police activity; the call for service and dispatch of a squad, the interview with the victim and followup investigation, the arrest and incarceration of a suspect, and the trial and disposition of the court case.)

References

Dangermond, Jack (1987), "PC/ARC/INFO Starter Kit—Part I," Video, Environmental Systems Research Institute, Redlands, Calif.
Koester, Edwin H. and Mercer, James L. (1978), *Public Management Systems*, AMACOM, a division of American Management Association, New York.
Martin, James and McClure, Carma (1983), *Software Maintenance*, Prentice-Hall, Inc., Englewood Cliffs, N.J.
Municipality of Burnaby (1986), "Invitation to Information," (brochure).
Naisbitt, John (1982), *Megatrends*, Warner Books, Inc., New York.
Strassmann, Paul A. (1984), *Information Payoff*, The Free Press, New York.
Wisconsin Land Records Committee (1987), "Final Report of the Wisconsin Land Records Committee: Modernizing Wisconsin's Land Records."

ADDITIONAL READINGS

Martin, James (1976), *Principles of Data-Base Management*, Prentice-Hall, Inc., Englewood Cliffs, N.J.

2

Geographic Information Systems Defined

> Essentially, all these disciplines are attempting the same sort of operation—namely to develop a powerful *set of tools* for collecting, sorting, retrieving at will, transforming, and displaying spatial data from the real world for a particular set of purposes.
>
> BURROUGH, p. 6*

The description of a geographic information system presented by Burrough acknowledges two significant factors that influence its definition. First, there are as many definitions as there are disciplines involved in using geographic information systems: geography, urban planning, engineering, data processing, landscape architecture, environmental science, and others. Second, a geographic information system consists of a set of tools that professionals in these disciplines use to improve the way they work. Just as telephones, calculators, and word processors are tools for making work easier, faster, and more meaningful, a geographic information system provides tools to professionals to improve their efficiency and effectiveness in working with map information and nongraphics attribute data.

Digital Map Information

It is important to distinguish between "map information stored in digital form" and a "map." Geographic information systems contain map information stored in digital form in a data base. They can then plot or display needed information from that data base to produce a map. Just as a printed report is one of the products of any information system, a map is one of the products of a geographic information system. The data base of digital map information is created from many different types of maps (e.g., tax maps, engineering maps and drawings, census

*Reprinted by permission from *Principles of Geographical Information Systems for Land Resources Assessment*, by P.A. Burrough (1986), Oxford University Press, Oxford, U.K.

maps, administrative maps). Maps are created from these data by selecting the desired map information from the data base and plotting or displaying it as an output of the system.

Nongraphics Attribute Data

Attribute data are descriptive information stored in a data base about features that can be located on a map. Usually, these features are physical objects that can be seen (as a building or street pavement); however, many times the features cannot be seen even though they can be located (as a parcel of land or a census tract). Often, attribute data are processed in a GIS for things other than features—call them phenomena, or incidents. These are situations that happen at a location (as a crime or a completed survey or even an election return). In all of these cases, attribute information is collected about something related to a location on a map.

The logical structuring of these data requires a *spatial reference*, or location attribute, stored in each record of the administrative files or data bases. This spatial reference (a parcel number, an address, an intersection, a block number, neighborhood code, census tract number, district number, etc.) is then linked to coordinates defined by a *geodetic reference system* and recorded on a continuous *digital map* of the jurisdiction. Retrieval, manipulation, and display of the geographic and map data are enhanced by the definition of *topological relationships* of all points, lines, and areas contained in the map information.

This is all accomplished by a set of GIS tools that can be used to automate service-delivery procedures, summarize data for managers, and provide new techniques for analyzing data for management, planning, and policy-setting purposes. These tools include the following:

Automated mapping technology—The automated mapping tools in an urban geographic information system provide flexibility in manipulating map information.

Data-base management—Data-base management tools in an urban geographic information system provide flexibility in managing attribute data.

Land records information—Land records information provides the cartographic and attribute data needed by an urban geographic information system to accurately and completely record maps and location-related attribute data.

Topological data structures—Topological data structures provide explicit definitions of the spatial relationships among points, lines, and polygons.

Spatial analysis capabilities—Spatial analysis tools provide the capability to retrieve, manipulate, and display map and location-related attribute data.

A thorough understanding of these tools is necessary in order to apply geographic information systems technology to the spatial data needs of an urban government.

GIS, AM/FM, CADD, and Other Labels

An enhancement to automated mapping systems is the automated mapping and facilities management (AM/FM) system. AM/FM systems utilize a data-base capability to store additional information about the mapped objects (physical features such as water valves, gas mains, meters, transformers, etc.) and link those data to the map information, but generally do not include spatial analysis capabilities or topological data structures such as those found in geographic information systems.

Professionals involved in this new technology have for many years expended a considerable amount of energy in defining the many names given to this technology. [In addition to CADD, AM/FM, and GIS, other names that have been used include: land information systems (LIS), land records systems, mapping information management systems (MIMS), computer-aided mapping (CAM)]. Driven mainly by the vendors of these systems who seek to maintain a "state-of-the-art" image for their product and differentiate their product from the competition, they have developed a plethora of definitions in order to communicate among themselves. Following are two recently published definitions from the GIS/LIS '88 Proceedings:

Exler:*

GIS—. . . The purpose of a traditional GIS is first and foremost spatial analysis. Therefore, capabilities may have limited data capture and cartographic output. Capabilities of analyses typically support decision making for specific projects and/or limited geographic areas. The map data-base characteristics (accuracy, continuity, completeness, etc.) are typically appropriate for small-scale map output. Vector and raster data interfaces may be available. However, topology is usually the sole underlying data structure for spatial elements.

CAD—Computer-aided design (CAD) systems store spatial data as graphic information. These systems handle maps as drawings with little or no data continuity across map sheets. Many CAD capabilities are

*Reproduced with permission from *GIS/LIS '88 Proceedings,* © 1988, by the American Society for Photogrammetry and Remote Sensing, v. 2, p. 815.

important in mapping. For example, flexible and high-quality carto-graphic displays and output may be created with capabilities typically found in CAD systems. However, this technology does not incorporate spatial analysis or geographic data management functions. These systems are designed for drafting and design/modeling and are primarily used in this industry for data capture and map publishing.

AM/FM—Automated mapping/facilities management (AM/FM) systems combine a subset of the CAD system capabilities for interactive graphics, entry, and storage techniques with a data-base capability. The system's focus is converting manual maps and records into a digital data base for query, work-order processing, and facilities model depiction and management. AM/FM systems often lack spatial analysis capabilities.

Klein:*

AM—Automated mapping: The use of computer graphic software systems to create, edit, and manipulate cartographic data. The resulting computer-generated graphic image files are a compilation of "primitive" drafted features with no information-processing capabilities other than to generate displays of different combinations of map data, views, and scales.

AM/LIS—Automated mapping/land information system: An AM/LIS refers to an LIS that uses computers to produce, edit, archive, and analyze the pictorial aspect of the overall system.

CADD—Computer-aided drafting and design: Process wherein the user interacts with a visual image on a computer screen to create, modify, or manipulate drawings. The mathematical integrity of components within the resulting graphic model is adequate to support legal survey and engineering design processes essential in an AM/LIS.

GIS—Geographic information system: "A GIS is best defined as a system which uses a spatial data base to provide answers to queries of a geographic nature. . . . The generic GIS that can be viewed as a number of specialized spatial routines laid over a standard relational data base management system." (Goodchild, 1985)

Clearly, the most distinguishing term used in defining a geographic information system is *spatial*. Definitions of CAD and CADD emphasize "mathematical" and "geometric" operations that are used in a much

*Reproduced, with permission, from *GIS/LIS '88 Proceedings*, © 1988, by the American Society for Photogrammetry and Remote Sensing, v. 2, p. 551.

larger set of applications beyond mapping, and, when used in a mapping environment, CAD and AM systems treat a map as an individual product or file, rather than a continuous geographic data base. The definitions of AM/FM, LIS, and MIMS systems refer to a map as a continuous geographic data base and include nongraphic (attribute) data bases for facilities, features, etc. used primarily as inventories of objects being mapped with limited analysis features. The most distinguishing term used in defining these systems is *mapping information.* The reference to *spatial* and *analysis* most often applied to GIS implies not only an ability to map information and refer to features that can be located, but also to identify relationships among mapped features and process their geometric characteristics for analyzing data in a spatial context.

This unique characteristic is the foundation of the definition adopted by the National Science Foundation, which created the National Center for Geographic Information and Analysis (NCGIA) in 1988. The prospectus for the NCGIA includes a definition of geographic information systems that is rapidly becoming the standard among GIS researchers and professionals across the country:

> A geographic information system is a computerized data base management system for capture, storage, retrieval, analysis, and display of spatial (locationally defined) data.

Automated Mapping Technology

Automated mapping technology consists of hardware and software that allow the user to convert, maintain, and display digital map information. Because of the large amount of data needed to represent digital map information and the complex processing involved in maintaining the data, these computer systems are usually separate from an agency's administrative information systems. (They may, however, be networked to these computers for the purpose of transferring data.) Most geographic information systems are purchased from vendors who "package" the various hardware and software items manufactured by the original equipment manufacturer (OEM) suppliers into a complete system that is tailored by the vendor to achieve certain performance characteristics. These vendors then sell their systems as complete packages including hardware, software, application programs, training, installation, and maintenance.

Hardware

The main feature of any computer system, including geographic information systems, is the *central processing unit* (CPU), also known as the central processor. The CPU of a geographic information system is no different from the CPU of other general-purpose information systems, and most of the popular CPUs are used in the packages provided by GIS vendors: microcomputers, minicomputers, and mainframes. Generally, however, the type of CPU needed in a geographic information system is based upon powerful and accurate 32-bit computer architecture, which refers to the amount of data the computer's logic circuits can transfer or process at one time. Since most mapping functions process large quantities of data, sometimes performing millions of calculations per second, the older 8-bit and 16-bit CPUs limit response time significantly. In addition, 32-bit CPUs offer twice as much precision as the 16-bit CPUs, which is important in mapping applications, in which accuracy and precision are critical.

There are two types of *magnetic storage devices* common to geographic information systems (which, again, are common to all computer systems): *disk drives* and *tape drives.* Both types of media (disks and tapes) are used to store digital data as well as application programs and other software. *Disks* provide random access, allowing immediate access to the data or programs (known as "on-line" storage). They are available in a variety of sizes, with storage capacities of less than 100 million characters (100 megabytes) to over 500 million characters (500 megabytes). They also are available in different physical characteristics such as "removable," which allows the physical replacement of one disk for another (used for copying data and programs as "backup" in case the on-line data is lost); "fixed," which is more reliable, since there is no physical handling of the disk by humans; and "floppy," which can be used to transfer data to other locations where a communication line is not available for such transfer. *Tapes* provide "off-line" storage of data and programs and are most often used as inexpensive permanent storage for archival purposes, transfer of data between different computers that are not networked together, and for easy transportation of programs and software upgrades from the vendor.

The *work station* provides access to the system by a user. Since there are different types of users (drafters, programmers, analysts, designers, operators), there are different types of work stations available. *Graphic work stations* consist of devices used to input, manipulate, edit, and display the data. The most unique devices in a graphic work station are the "digitizing" board or table (sometimes called a pad or tablet) and its associated "mouse," "digitizer," or "puck." The mouse on the digitizing table allows the user to select a point from a map or drawing and automatically enter its *XY* coordinates to the computer. With sizes ranging from 8 inches by 8 inches for a graphics pad to over 40 inches square for a digitizing table, most maps and engineering drawings common to local government can be accommodated. The electronic configuration of the board and the mouse differ among the various vendors of these devices, but the concept of how they automatically assign *XY* coordinates is simple: Tiny wires (or electronic impulses) run horizontally and vertically inside the board, about 0.005

inches apart, forming a grid. When the mouse is placed on the board, usually with a map or drawing between the surface of the board and the mouse, an optical viewer with crosshairs on the mouse allows the user to locate visually a point on the map or drawing. When that point is identified, the user pushes one of several buttons on the mouse and an electrical impulse is triggered. Since the mouse is connected to the computer with a wire, and since the digitizing table is also connected to the computer, the exact location of the impulse in relation to the nearest intersection of the tiny horizontal and vertical lines is known (within 0.005 inch, of course). This allows the computer to translate the location of the point on the table into the location represented on the map (because the map has been registered on the table previously by the user with different buttons on the mouse and functions of the computer). Since most systems are "interactive," the location of the mouse on the table (and, hence, the map) is simultaneously displayed on the graphic display screen. This process is called *digitizing*—converting hardcopy maps into digital form for use by the computer. In addition to digitizing, the mouse is also used to select points on the display screen, select commands from menus on the screen or on the table, draw freehand without a map or drawing on the table, and move windows around on the screen—all functions being performed by depressing the appropriate button on the mouse in conjunction with the display on the screen.

The *graphic display screen* is a high-resolution cathode ray tube (CRT) used to display the graphics representation of the map. It is high resolution because it displays graphic features by turning on individual dots on the screen (called *pixels* because they are *pic*ture *el*ements) that are very close together—much closer than those in alphanumeric terminals. High-resolution graphics display screens generally provide a pixel density of 1000 by 1200 points on the screen. (Compare to the pixel density of 960 by 532 points for alphanumeric screens. A common television, by the way, provides a pixel density of 512 by 512 dots on the screen.) These graphics display terminals are called *raster displays* because they display lines as combinations of dots very close together. Older graphics display terminals were *vector displays*, which actually drew continuous lines on the screen, but also required constant "refreshing" by the CPU and thus used up substantial processing time, which affected response time. Graphics display terminals are available in color or monochrome (one color with black or white) models. The most expensive models can display up to one million different colors.

The final device found in most graphics work stations is the *keyboard*, which is similar to keyboards found in most computer systems. The keyboard allows entry of data or commands to the computer (often the same commands as the mouse) and usually has function keys to perform a series of commonly used commands merely by depressing a key. Keyboards are a convenient method for entering alphanumeric data during the digitizing process.

Programmers and data analysts often do not require the expensive graphics capabilities of the graphics work station. When used for programming new applications, retrieving attribute data for tabular output, and for computer operations functions, a simple and inexpensive alphanumeric terminal and keyboard is used as a work station.

Many work stations contain microprocessors that allow the work station to operate as its own computer, thus reducing the demands on the main computer. Typically, data files are transferred to these "intelligent" work stations, and map and attribute data are manipulated by the software in the work station. If the data have been updated or otherwise changed, then the files are transferred back to the main computer for updating the data bases so that other work stations may have access to the updated data.

The *plotter* is another hardware device that distinguishes automated mapping systems from general-purpose computer systems. Plotters are output devices that produce maps and drawings in "hard copy" form. They can produce these outputs either by using a pen, as with a *pen plotter*, or through an electrostatic process, as with an *electrostatic plotter*. Both types of plotters produce high-quality maps and drawings. There are many different types of pen plotters: Sizes range from 8 inches by 10 inches to 4 feet by 8 feet, with the most common being about 40 inches by 40 inches. The number of pens also varies, and the plotting methodology varies from a "flatbed plotter," in which the paper is flat and the pens are moved on it, to a "drum plotter," in which the paper lies on a drum mechanism that moves in a vertical direction while the pen moves in a horizontal direction. A variety of media can be used with pen plotters, including paper, vellum, and Mylar. Pens also vary, including ball point, felt tip, and liquid ink. Because of the variety of pens available, color plots can be produced on pen plotters.

Electrostatic plotters are quite different because they use raster technology, as described in the previous discussion on graphics display screens. Electrostatic plotters do not use pens to apply the ink to the paper; they use thousands of tiny needles all in one row across the paper (densities vary from 100 to 400 per inch). When the paper is transported over the needles by a drum mechanism, each needle transfers a charge to the paper if ink is to be placed at that location on the paper. If no ink is to be placed at the location of the needle, no charge is transferred. As the paper continues past the needles, ink is transferred to the charged dots on the paper. Because the dots are so close together, a line appears solid as if a pen were used. Since the computer, however, stores the map information as vectors (a line with a beginning point, a distance, and a direction) instead of dots as used in the raster output, the map information must be translated from vector format to raster format prior to plotting on the electrostatic plotter. Thus, another device, a *vector-to-raster converter*, is needed in conjunction with the electrostatic plotter.

Electrostatic plotters are also available in different sizes and can produce color plots. While they are more expensive than pen plotters, they are much faster in producing plots (60–100 times as fast). Other advantages of electrostatic plotters over pen plotters are: Shading of polygons can be more continuous because pen plotters must use lines rather than dots, and electrostatic plotters do not require constant pen replacement when different line weights are used on a map or drawing.

These hardware devices: the CPU, disk and tape drives, work stations, and plotters (see Figure 2.1) are common configurations found in most automated mapping systems and geographic information systems. There are, however, additional devices used in these systems for special applications that are not universal for all local governments. Three of these additional devices will now be described.

Optical scanners are a relatively new technology that someday, one hopes, will replace digitizing for converting map data into digital form. Digitizing, the reader will recall, is the process of identifying points on a map using the mouse and digitizing table. It is now the most com-

FIGURE 2.1. Hardware devices in a geographic information system.

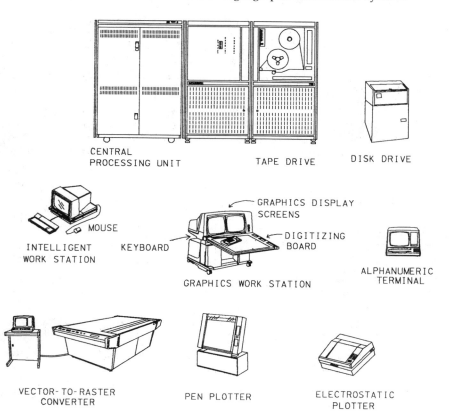

mon method for converting existing maps (some jurisdictions, however, choose not to use existing maps but, instead, create new ones from aerial photography or coordinate geometry software). While it is a common process, however, it is also very tedious when converting high-density parcel maps of local government. (A typical "quarter-section map" ½ mile by ½ mile square with individual parcels can require up to 80 hours of work to digitize—including checking and correcting errors.) Optical scanners, in some situations, can replace digitizing by automatically converting hard-copy maps to digital form. They do this through a "raster-to-vector" process that is similar to the vector-to-raster process described earlier for producing electrostatic plots.

The raster-to-vector process of optical scanning is the use of a laser beam to scan the surface of a map and convert it to thousands of tiny dots that are coded as either black or white, depending upon the presence of ink on the map. These dots are stored as a "raster image," a matrix of black and white dots, which has a dot for every possible location on the map. Since this matrix requires a very large amount of storage, a more efficient storage method is used to store the map information in the computer: combining all the raster dots that make up a line into a vector, which requires only a beginning point, a length, and a bearing for storage. Thus the optical scanner reads the map, converts it to a raster image, and then combines adjacent dots into vectors. This process cannot be accomplished, however, without a human helping to identify which dots are associated with others, and which are superfluous. Since the scanner picks up all marks on the map and converts them to the raster image, it will convert coffee stains, dirt, and other marks that should not be part of the map. These must be identified and eliminated by the human operator prior to vectorizing the image. In addition, all text symbols (letters, numbers, words, and special characters) are also scanned as raster images and therefore cannot be stored in the computer as data items (vendors of optical scanners are now developing methods for converting this text information directly into data formats, but have yet to perfect the capability). For now, a human operator must assist in the capture of text information during the map scanning process. This "cleaning" process can consume substantial operator time, especially when the maps have a significant amount of nongraphic information (including dirt, etc.) that should not be vectorized. With today's technology, optical scanning of maps characteristic of local government can require more time to convert to digital form than digitizing the maps.

Computer output microfilm (COM) is additional hardware that can produce a map or drawing directly onto microfilm. It is a device that creates an electronic beam that draws the map on the film, which is later developed to produce the microfilm that is generally stored on an aperture card for producing white prints upon demand. COM units are expensive ($50,000 to $100,000) but can be justified in agencies having a high volume of map maintenance and reproduction costs. The use of COM can eliminate the need for hard-copy plots of standard map products when they are normally microfilmed through a photographic process for later reproduction.

Agencies that require contour lines for mapping elevations of the land often use *stereo digitizing plotters* as input devices to automated mapping systems. These devices are used to digitize elevation data as Z coordinates from stereo pairs of photographs when viewed by an operator through a viewing device that gives a three-dimensional view of the photograph. By identifying physical markers on the photograph that have known XY coordinates and elevations, the coordinates and elevations of all other significant points on the photograph can be recorded.

Software

Automated mapping software consists of interactive computer graphics programs that can create, edit, manipulate, and display cartographic data. These programs are similar to those found in computer-aided design and drafting systems because they allow the user to interact with a visual image of a drawing by creating, editing, and manipulating lines, symbols, and text. Automated mapping software generally has the same functions as CADD software; however, CADD systems are normally used for architectural and engineering drawings, while automated mapping is used for mapping. Functions specific to mapping include: coordinate transformation, map scale conversion, coordinate geometry (COGO), edge-matching, and other related geometric operations. Software for performing spatial analysis will be described in this chapter.

Coordinate geometry software translates alphanumeric data input at a keyboard (coordinates, distances, bearings, and other legal descriptions of property) into digital map information for creating and updating a cartographic data base. It is a set of mathematical programs that translate survey information into map information and is used by some agencies that do not have property maps or whose property maps lack appropriate accuracy or quality for digitizing.

Edge-matching programs manipulate features at the edge of adjacent map sheets and fit them logically into continuous features across map sheet boundaries. Manual map records consisting of individual map sheets bound together into an atlas do not use a continuous coordinate system and therefore may require edge-matching to make them "fit" together when digitized. Thus, edge-matching is used to align street centerlines, curb lines, water and sewer mains, etc. into continuous lines across the entire jurisdiction.

Coordinate transformations are programs that mathematically convert coordinates from one frame of reference to coordinates of another frame of reference. Thus, maps digitized on a local coordinate system can have the coordinates of all of their features translated to state plane coordinates or latitude and longitude coordinates by using coordinate transformation programs.

Windowing (also known as "zooming") is a simple function that allows the user to identify a smaller geographic area than that which is displayed on a screen and have it enlarged on the same or adjacent screen. It is an electronic magnifying glass that also retains the intelligence of the map by maintaining the logical associations of the attributes, regardless of the scale displayed.

Curve fitting (also known as "coordinate filtering" and "thinning") is a function to convert a series of short, connected straight lines into smooth curves to represent features that do not have precise mathematical definitions (such as rivers, shorelines, and contour lines). The more short lines digitized, the smoother the curve is.

Area calculation computes the area of a polygon when the polygon is defined as a shape with boundary lines that close the polygon. Perimeters are also calculated from polygon definitions.

Line-length calculations compute the length of a line identified by the user. The length of a line that consists of more than one segment, such as a curved line, can be calculated by identifying the beginning point and the ending point of the entire line.

Scale conversion is the mathematical conversion of a map from one scale to another. Scales themselves can be converted (from 1 inch = 100 feet to 1 inch = 400 feet), as well as the entire system of measurement (from English to metric).

Text placement allows the placement of letters, words, and numbers at a specific point identified by the user.

Snapping makes two lines meet mathematically during the digitizing process. This is used to ensure connectivity between lines when it cannot be visually verified.

Copy parallel allows the user to identify one line and have it copied a certain distance, creating two parallel lines.

Data-Base Management Systems

Geographic information systems are, first and foremost, information management systems. They help integrate data collected by the many different functions of local government about features on the ground, in the buildings, and under the ground. This integration occurs not only between the map and the attribute data, but also between organizational units. Fire personnel need to know the owner of a burning building. Tax assessors must know what permits have been issued for alterations to buildings. Sanitation managers need to know how many dwelling units there are in blocks on which garbage routes need adjustment. Utility companies need to know what public facilities are under the ground where they are about to dig. These are information needs that cross functional areas (Fire and Tax, Tax and Building Inspection, Sanitation and Tax, Utility and Public Works). They are also needs that relate to the same location (an address, a block, a street segment). They are horizontal data-integration needs of local government that can be satisfied only by effective data management.

As described in Chapter 1, traditional information systems—transaction-based systems—are not easily adapted for such multiple purposes because they were originally designed for a single purpose. The computer programs in these systems are too inflexible to modify for other uses and the data files they use, too difficult to access by different means. Data-based management systems, however, separate the data from the programs that access it, reducing these barriers and enhancing horizontal and vertical data integration.

The system described in Chapter 1 that automated the process of identifying building code violations and printing violation notices was not a data-based system. It was a transaction-based system that read marks on an inspection card after the inspector found violations at a building and then automatically printed a code violation notice that was sent to the building owner the next day. It was a smashing success because it reduced the department's need for staff from six typists to two clerks. It improved productivity and reduced the cost of government.

Then the head of the department retired, and the new leader replaced some key managers. These new managers, after studying the operation of the system, felt that they should also be able to use the computerized data to list all of the buildings with violations that were not abated (corrected). These lists, printed on a weekly or monthly basis, could be used as work orders to remind the inspectors which buildings required reinspecting. (The new managers had observed that the office copies of many of the notices were torn, spilled with coffee, forgotten under the seats of the automobiles, and even lost in the paper shuffle!) Furthermore, if the lists could be sorted by inspection district, then the supervisors could use them to evaluate the performance and workload balance of the inspectors.

Next, one of the city's aldermen, concerned about the city's poor housing conditions, decided to launch an attack on the landlords who owned many properties having a large number of building code violations. He asked for a list of the top 12 landlords whose properties had the most code violations: the "Dirty Dozen."

Then the city's Planning Department, studying the effects of neighborhood deterioration, asked for summary information by city block of houses that had serious defects so that they could establish boundaries of areas designated for low-interest home improvement loans.

Because of the need for these additional applications, the system was redesigned from an efficiency-oriented, transaction-based system (essentially a glorified typewriter) to an effectiveness-oriented, data-based system. A data-base management system was used to transform the computerized inspection system into a flexible, integrated information system that not only *crossed organizational boundaries* (Building Inspection, Common Council, and Planning), but also was *linked to the geography* (address, city block, inspection district) and was *linked to other data* (property ownership records). The new applications could not have been implemented as modifications to the original system because their needs evolved over time and were not known at the time the system was designed. Once the system design was based upon the structure of the data (rather than the processing of notices), it was flexible enough to accommodate additional uses without requiring major changes.

Data-Base Structures

An urban geographic information system must be able to process a variety of attribute data from many different sources in order to satisfy the information needs of the service delivery, management, and policy levels of government. Much of the information needed by officials in those levels has already been computerized in existing transaction-based information systems (such as the building-inspection example discussed previously), and the task of the GIS professional is to integrate these systems, or at least their data bases, with the new technology. In order to accomplish this integration, it is necessary to understand the different methodologies currently used to store attribute data in today's information systems because the logical structuring of the data determines the degree of flexibility of the systems.

FLAT FILES

Traditional information systems without data-base management software process data structured in the form of "flat files." These data files are called "flat" because each record in the file contains the same data items (also called "fields") as the other records (Martin, 1976, p. 85). The values of the data items, or fields, differ, but the items themselves are the same. Usually, one of the items is designated as a "key" field, which is used for locating a particular record or for sorting the file in a particular order.

The Property File example shown in Figure 2.2 is a flat file containing four records, each record having the same ten data items. To find the assessed value of one of the parcels, a program must find the Parcel Number (the key for this file) for the parcel of interest and then print the contents of the Value field of the record for that parcel.

Since most local governments in urban areas have very large data bases containing thousands, ten of thousands, and even hundreds of thousands of records in each computer file they maintain, how computer programs search these files for the desired data is the essence of flexibility and responsiveness in urban data management.

Sequential search—Flat files are usually ordered sequentially by the values in a field designated as a key. Searching for a record based on the value of its key is fast because, once the desired record is found, the search can cease. This would make a parcel number search on the Property File fast because Parcel Number is a key; however, the entire file must be resequenced in order to find the records with a particular owner (twice, in fact—once for Owner 1 Name and once for Owner 2 Name). Any search other than on a key field requires resequencing the entire file or else every record in the file must be read by the program.

Binary search—Files sequenced by the values in the key field can be searched much more rapidly by using a binary search technique that starts at the middle of the file rather than at the beginning. It compares the value of the key in that record with the desired value for a "greater than" or "less than" condition and eliminates for consideration the half of the file in which the desired record could not be stored (because the file is sequenced by the values in the key field). It then goes halfway again through the remaining half and compares the value of the key field in that record to the desired value, eliminating half of those records (now one-quarter of the original file). It continues halving the records for consideration until the desired record is found. Again, the entire file must be resequenced if a different field is to be used as a key (such as Owner Name).

Indexed search—Whether the file is sequenced or not, sometimes an index is used to find a desired record. An index is a separate table containing the key of every record and an address pointing to the location of the data record for each key. The address is the location on the physical storage device used and is assigned by the software that controls the device. The index is then sequenced, and a search (usually binary) is used on the index rather than on the data file. This is much faster than other searches because the size of the table is smaller than the data file. Additionally, if a new record is added to the file it can be added at the end without resequencing the entire file—only the index need be resequenced. Think of an indexed search as looking for a book in the library. The card index is used to find the catalog number of a book with a certain title. The catalog number provides the actual location of the book. Just as the library has more than one card index (one for title, one for author, and one for subject), so can a data file have more than one index (such as Parcel Number and Owner Name in our example). Indexed files allow access by more than one key; however, they require a considerable amount of overhead to keep the indices updated when records are added and deleted, and when values in the records change over time.

While some data-base management systems use flat file data structures for storing data, their capabilities are severely limited and offer little flexibility. Flat file structures are simple and efficient for specific, repetitive tasks such as found in transaction-based information systems; however, they present a number of problems when flexibility of use and responsiveness are required:

- Access to records by any field other than the key field is very slow;
- Expansion of the records to accommodate additional fields requires major reprogramming work;
- Adding new records to the data file requires additional processing tasks.

HIERARCHICAL FILES

In a hierarchical file structure, there is more than one type of record in the data base. One type of record is designated as a "parent" or master record, and it can be associated with any number of "children" or detail records through internally assigned pointers. The detail records can also have "children" associated with them, with additional pointers assigned for the third level of association. The distinguishing characteristic of these hierarchical data structures is that each record has one higher-level record associated to it. The advantage with this type of structure is that it allows multiple sets of like attributes to be associated with any given record, without storing the repetitive data all in one file. This is generally known as a "one-to-many" relationship.

The flat Property File (Figure 2.2), for example, has one record for each parcel and contains room for storing the names and addresses of two owners of the parcel. Data storage is wasted when there is only one owner, and space is not available for more than two owners. Adding more room to the file for recording the additional owners causes a major change to the file (adding two more fields: Owner 3 Name and Owner 3 Address). This restructuring of the file would affect all programs using the file, requiring them to be modified (even if they do not use the new field).

Figure 2.3 shows the same data in a hierarchical structure. Here, there are two types of records: a Parcel Master Record and an Owner Detail Record. The data are the same as the data in the flat file, except that each parcel can now have any number of owners associated with it (see Parcel 075). The linking of each parcel record to the owner records associated with it is accomplished by pointers assigned by the data-base management software when each detail record is created. Any time a new owner record is created, the data can be added anywhere in the file because the software automatically links the location of the data in the file to their associated parcel record.

While the hierarchical data structure provides some flexibility in relating data between records, there are some serious limitations in using hierarchical files in a data-base management system:

PROPERTY FILE

Parcel Number	Parcel Address	Block	District	Tract	Owner #1 Name	Owner #1 Address	Owner #2 Name	Owner #2 Address	Value
008	501 N SADOWSKI ST	1	A	101	SADOWSKI, M.G.	501 N SADOWSKI ST			105450
009	590 N SADOWSKI ST	2	B	101	ADAMS, JULIE A	590 N SADOWSKI ST	ADAMS, M	590 N SADOWSKI ST	89780
036	1001 W ADNAN RD	4	B	105	SADOWSKI, M.G.	501 N SADOWSKI ST			101500
075	1175 W DADLEZ DR	12	E	202	KROEGER, ROSS	592 N TIERNEY PL	BERTRAND, K	1087 W BERTRAND DR	98000

KEY FIELD

FIGURE 2.2. A flat file structure for a property records data base. There is one record for each parcel, and all records are the same length. Access to the data in a record is accomplished by searching through all records for a value in one field, called a key field. Here, the key field is Parcel Number.

PROPERTY DATA BASE

PARCEL MASTER RECORD

Parcel Number	Parcel Address	Block	District	Tract	Value
008	501 N SADOWSKI ST	1	A	101	105450
009	590 N SADOWSKI ST	2	B	101	89780
036	1001 W ADNAN RD	4	B	105	101500
075	1175 W DADLEZ DR	12	E	202	98000

POINTERS

OWNER DETAIL RECORD

Parcel Number	Owner Name	Owner Address
008	SADOWSKI, M.G.	501 N SADOWSKI ST
009	ADAMS, JULIE A	590 N SADOWSKI ST
009	ADAMS, M	590 N SADOWSKI ST
036	SADOWSKI, M.G.	501 N SADOWSKI ST
075	BERTRAND, K	1087 W BERTRAND DR
075	KROEGER, ROSS	592 N TIERNEY PL
075	KROEGER, S	592 N TIERNEY PL

FIGURE 2.3. A hierarchial file structure for a property records data base. There is more than one type of record for each property (Parcel Master and Owner Detail) with pointers allowing a "one-to-many" relationship where each parcel master record can have one or more Owner Detail records associated with it. Access to the data is limited to one type of record (the "Parent" record).

41

- Data in the detail records can only be accessed by first accessing the master record to which they are linked (the property data base in Figure 2.3 cannot be accessed by Owner Name);
- Data in the detail records must be repeated for each master record association (Owner Address must be repeated for each property owned by a given owner).

NETWORKS

When detail records can be associated with more than one master record, the structure is defined as a *network*. The network is established by using pointers that link records between related files. This "many-to-many" relationship allows more than one record in a file to be associated with more than one record in another file. For example, the property data base contains records of parcel data, and some parcels are owned by more than one person. Any one person, however, can own more than one parcel. This is a "many-to-many" relationship, which network data structures accommodate better than hierarchical structures.

Figure 2.4 represents the property data base in a network structure. The network consists of four files containing different, yet related, records. They are related through key fields that are common between the files: Parcel Number and Owner Name. Pointers link the records of the different files that are related through these fields. Thus, Parcel 075 in the Parcel Master File is linked by three pointers to the three owner names in the Parcel Detail File. The names, in turn, are linked to the Owner Master File, which contains the addresses of the owners. (Notice that the address of each owner is recorded only once—in the Owner Master File. All other parcels owned by a certain owner are linked to this record by pointers so that Owner Address need be recorded only once.) Using a different network path, all parcels owned by "Sadowski, M.G." can be found by first searching the Owner Detail File on that name and identifying the Parcel Numbers associated with "Sadowski, M.G." (Parcels 008 and 036). These records then point to the records in the Parcel Master File containing the same Parcel Numbers, giving the addresses of those parcels.

This is the nature of networked files—pointers between related records of different files. If an owner moves, then only one record—that in the Owner Master File—must be changed. If a parcel is sold to a different owner, then only the Parcel Detail File and Owner Detail File need updating. If a new parcel is created on the Parcel Master File, then a new record must be created on the Parcel Detail and Owner Detail Files, and, if the owner is new, then a new record must be created on the Owner Master File. While all these changes are occurring, the DBMS software is keeping track of the pointers, ensuring that the network relationships are intact.

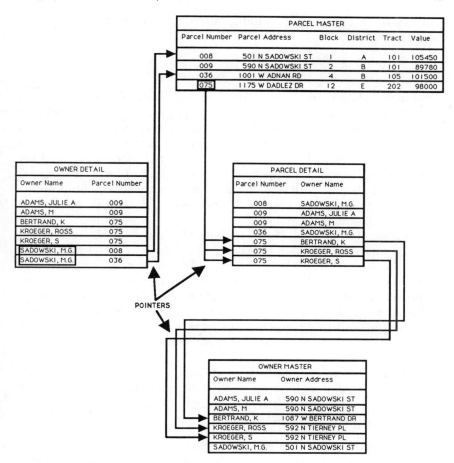

PARCEL MASTER					
Parcel Number	Parcel Address	Block	District	Tract	Value
008	501 N SADOWSKI ST	1	A	101	105450
009	590 N SADOWSKI ST	2	B	101	89780
036	1001 W ADNAN RD	4	B	105	101500
075	1175 W DADLEZ DR	12	E	202	98000

OWNER DETAIL	
Owner Name	Parcel Number
ADAMS, JULIE A	009
ADAMS, M	009
BERTRAND, K	075
KROEGER, ROSS	075
KROEGER, S	075
SADOWSKI, M.G.	008
SADOWSKI, M.G.	036

PARCEL DETAIL	
Parcel Number	Owner Name
008	SADOWSKI, M.G.
009	ADAMS, JULIE A
009	ADAMS, M
036	SADOWSKI, M.G.
075	BERTRAND, K
075	KROEGER, ROSS
075	KROEGER, S

POINTERS

OWNER MASTER	
Owner Name	Owner Address
ADAMS, JULIE A	590 N SADOWSKI ST
ADAMS, M	590 N SADOWSKI ST
BERTRAND, K	1087 W BERTRAND DR
KROEGER, ROSS	592 N TIERNEY PL
KROEGER, S	592 N TIERNEY PL
SADOWSKI, M.G.	501 N SADOWSKI ST

FIGURE 2.4. A network file structure for a property records data base. There is more than one type of record with pointers allowing related records to be associated with each other in a "many-to-many" relationship. Access to the data can be made on any type of record, and pointers are used to associate different records with related data.

As the use of a data base expands by adding new applications and by providing responses to new types of data inquiries, the logical linkages among data fields tend to multiply, causing a high degree of complexity in managing the pointers within the network. In complex networks with many logical relationships and very large data bases, the amount of storage required to store the pointers can exceed the storage of the data base itself. The management of these pointers, as records are added and deleted, as new fields and linkages are created, and as data values change, can become so cumbersome that the data-base system can be in danger of becoming inflexible and unresponsive to the needs of the user.

RELATIONAL DATA BASES

A relational data base allows related records from different files to be associated with each other without using pointers or keys to represent the linkages among the files. This reduces the complexity of the network by allowing the logical linkages of the data values in common fields among files to form the associations. This process is called *normalization* and consists of the creation of separate files with common logical linkages (common data items) to replace the associations among data items that are represented by pointers and keys in hierarchical and network structures.

A relational data base, then, consists of a series of files called *relations*, or tables. Each table contains a data item—a column of data—that is the same as one or more other tables containing additional data. The two (or more) tables are related by this common data item (column of the table) and then can be "joined" to form a new table. This joining is similar to pasting columns of printed tables to produce a new table. (A table can also be "projected" to form a new table of only certain columns from the original table with values sorted in any desired order.)

The major advantage of this type of data-base structure is the almost unlimited flexibility in forming relationships among data items in the data base without suffering an undesirable degree of linkage management. As Martin (1977) describes it:

> Different users of the same data will perceive different sets of data items and different relationships between them. It is therefore necessary to extract subsets of the table columns for some users, creating tables of smaller degree, and to join tables together for other users, creating tables of larger degree. . . . These cutting and pasting operations give a degree of flexibility that is not possible with most tree (hierarchical) structures and plex (network) structures.*

Without keys and pointers to manage, new relationships between data items can be created "on the fly" by the user rather than depending upon a computer programming specialist to establish the associations.

Figure 2.5 shows a relational data-base structure for property information. The four tables are related by Parcel Number, Owner Name, and Block. By relating the Parcel Table and the Owner Index Table on their common Parcel Number column, a new table of all parcels and their owners is created as shown in Figure 2.6. When the Parcel Table

---

*James Martin, *Computer Data-Base Organization*, second edition, © 1977, p. 204. Reprinted by permission of Prentice Hall, Inc., Englewood Cliffs, New Jersey.

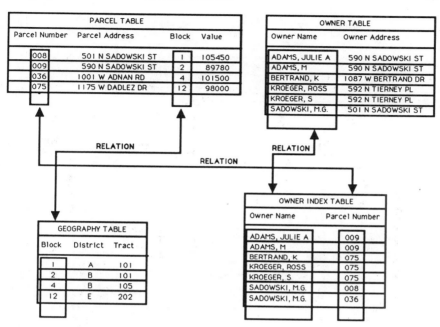

PARCEL TABLE

Parcel Number	Parcel Address	Block	Value
008	501 N SADOWSKI ST	1	105450
009	590 N SADOWSKI ST	2	89780
036	1001 W ADNAN RD	4	101500
075	1175 W DADLEZ DR	12	98000

OWNER TABLE

Owner Name	Owner Address
ADAMS, JULIE A	590 N SADOWSKI ST
ADAMS, M	590 N SADOWSKI ST
BERTRAND, K	1087 W BERTRAND DR
KROEGER, ROSS	592 N TIERNEY PL
KROEGER, S	592 N TIERNEY PL
SADOWSKI, M.G.	501 N SADOWSKI ST

RELATION   RELATION

RELATION

OWNER INDEX TABLE

Owner Name	Parcel Number
ADAMS, JULIE A	009
ADAMS, M	009
BERTRAND, K	075
KROEGER, ROSS	075
KROEGER, S	075
SADOWSKI, M.G.	008
SADOWSKI, M.G.	036

GEOGRAPHY TABLE

Block	District	Tract
1	A	101
2	B	101
4	B	105
12	E	202

FIGURE 2.5. A relational file structure for a property records data base. Here, four separate flat files (called *relations*) contain related data that can be combined by matching records having the same values in columns common to the files.

is related to the Geography Table by Block, we can obtain a table of all parcels with their geography and assessed value (see Figure 2.7).

Summary

A geographic information system does not consist of a single data base for all of its applications. The extensive investment that local governments have made in computerizing their data over the past 30 years prohibits a complete restructuring of all of their data to create a single data base for an urban geographic information system. As long as their data content and structure provide for a geographic linkage, or key, among the files, then there is no need to combine all location-related data into one common data base. To do so would create havoc with the management of pointers and indices and may adversely affect the nongraphics applications for which they were built.

In addition, the foregoing discussion addressed the structures for only the attribute data processed in a GIS and not the structures for the cartographic data—the map information. Generally, digital map data are not managed by a system's data-base management software, but rather through the proprietary design of each vendor's product. This

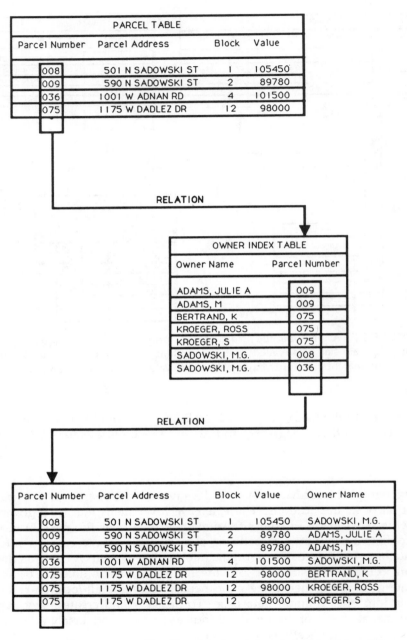

FIGURE 2.6. Joining two tables in a relational data base. A table with parcel information (Parcel Table) is "joined" with a table of owner names (Owner Index Table) to create a new relation with parcel information and owner information.

is because no standards have been imposed on the GIS vendors (such as the CODASYL standards for DBMS mentioned in Chapter 1) and because cartographic data are different from attribute data. The rules are different. Cartographic data are much more interrelated than attribute data. As will be seen later in this chapter, a change in the *xy* coordinate of a single point can cause a substantial ripple effect of changes to all other cartographic elements related to that point (lines, polygons, text, etc.). The disadvantages listed in the summary of characteristics of the different structures (Table 2.1) include numerous

FIGURE 2.7. Joining two tables in a relational data base. The Parcel Table is "joined" with a Geography Table to create a new relation of parcel information with geographic codes.

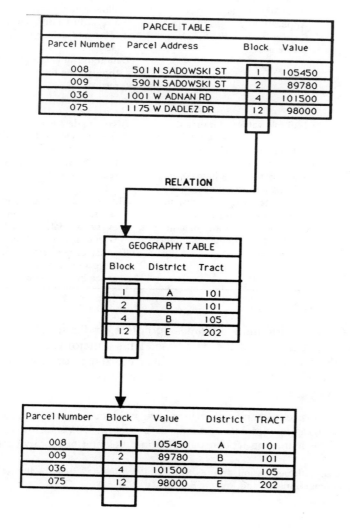

Table 2.1. A summary of data-base file structures.

	Flat	Hierarchical
Characteristics	All records contain the same data items Records are ordered sequentially by a data item designated as a key Records are accessed on the key either by sequential search, binary search, or by reference to an index All data items are functionally dependent on the key	Data items are contained in more than one record type Each record in a file is associated with one higher-level record in a different file Pointers are used to associate records of a lower level with their higher-level record
Advantages	Data retrieval is fast when using the key to find a record Applications are easy to program because the file structure is simple	Multiple sets of like records are allowed to be associated with a record of a different file Adding and deleting records are easy Data retrieval is fast if access is through the higher-level record
Disadvantages	Multiple values of a data item for a given record are difficult to process and complex to program Adding new data items or expanding the length of any data item requires substantial program modifications Data retrieval on any data item other than the key is cumbersome and slow	Access to data is restricted to the path established by the pointers (must first go through the higher-level record to get to its associated records) The same data in lower level records must be repeated for every association they have with a higher-level record The use of pointers can require large amounts of storage space in large data bases

problems with maintaining linkages (pointers, indices, relationships) between related records of the data base. Thus a major characteristic of cartographic data (association with other records) is one of the most disadvantageous characteristics of DBMS data structures. This problem is described by Dangermond (1988):*

It is important to recognize that there are fundamental differences between managing tabular data and map data. Present DBMS technology is very good for managing tabular data (adding, deleting or modifying rec-

*Reproduced, with permission, from *GIS/LIS '88 Proceedings*, © 1988, by the American Society for Photogrammetry and Remote Sensing, v. 2, p. 568.

Network	Relational
Data items are contained in more than one type of record	Data items are contained in different flat files called *tables*
Records can be associated with more than one record in other files	Each entry in a table is one data item—there are no repeating fields
Pointers are used to associate related records of different files	Tables can be related by the values in the columns that are common to the tables
	New applications are developed by forming new relations between tables
Access to records in different files is allowed in many different paths	Data are easy to access by nontechnical users
Like data items are not repeated in multiple associations	New uses and unanticipated inquiries can be easy to implement
Changes in the records of one file do not affect programs that use other files	Adding new records, new data items, and new relationships between records do not affect existing programs
Pointers defining associations between records of different files are automatically changed when records are added or deleted	The physical data storage can change without affecting the data or logical relationships between records
Expansion of use over time increases the complexity of the associations and can cause the system to become cumbersome and inflexible	New relations between large tables require substantial processing time
New associations are difficult to establish because the data must be restructured for the new associations	Access to records in a table is sequential and can be slow
Complex networks with large data bases require extensive storage for the pointers	How the data is stored physically on the disks has a significant impact on processing time
	The high degree of flexibility in relating records of different tables can allow logical mistakes to be made when the user forms new relations on invalid combinations of data items

ords), but mapping involves much more than just the storage and retrieval of data. DBMS technology is not very effective for updating and managing cartographic data, except the attribute data associated with spatial features. This is so because of the topological relationships between map features. When, for example, a land parcel's polygon is adjusted, the relations of surrounding parcels to the first parcel and to each other must also be adjusted; the geometric attributes of all these parcels change. Thus manipulating spatial data is much more complex than managing tabular data. So, coordinate and topological data fundamental to a GIS do not belong in the traditional tabular DBMS environment.

Land Records Information

Land records provide the basic references needed by an urban geographic information system to add meaning to the spatial and location-related data it processes. Attribute data on land parcels, buildings, census tracts, street segments, etc. can be manipulated and analyzed interminably with little meaning if the results cannot be related to physical locations on the earth, represented by a map. Before GIS technology was available, traditional information systems were used to process location-related data and produce reports in tabular form. These reports had meaning when they were used with a map as a reference (total registered voters by voting ward had meaning when the locations of voting wards were seen on a map; the number of traffic accidents per year by street intersection meant something when a street map was used to see where the intersections were located). Often, the attribute data listed and summarized in tabular outputs were manually transferred to maps for further review, display, and analysis. Now that GIS technology contains the map information in digital form, however, the attribute data and the map data can be processed together as a single product with little manual intervention.

To be able to process the map information together with location-related attribute data requires accurate, consistent, and meaningful land records as a base for reference. (A parcel map has restricted use if it is not placed in context with other parcels, the street network, and the jurisdiction as a whole.) It is not enough to have accurate map information—the maps themselves must be related to each other and to physical locations on the earth. In addition, the features that maps represent (intersections, parcels, voting wards, streets, etc.) must be identifiable on both the map data and the attribute data in order to relate the data to the locations of the features. (What good is knowing the name of the owner of a parcel if the parcel is not also identified—all you would have is a name!) Accurate, consistent, and meaningful land records provide the references necessary for a GIS to relate attribute data to map data and to locate that information in its actual physical location on the earth.

The references used by an urban geographic information system include a *uniform coordinate system* to display the cartographic data in the form of a map and *location identifiers* to process and display attribute data about features on the map. This is an important concept in a geographic information system. Merely storing cartographic data as a visual image of a map without placing it on a continuous coordinate system prevents its use across the agency's entire jurisdiction. Merely storing attribute data describing the features managed or affected by the agency does not allow them to be located on a map. The maps containing the cartographic information of the jurisdiction must be registered on a continuous coordinate system for the entire area or else they are just electronic map sheets, their use being limited just as pages of an atlas limit the use of hard-copy maps. Similarly, the attribute data bases containing descriptive data about the features must have location identifiers (addresses, parcel numbers, block numbers, census tract numbers, etc.) to link the data to the map, or else they can only be used to print lists and summary reports.

Figure 2.8 and Figure 2.9 portray this concept graphically. Figure 2.8 represents land records for a GIS application involving parcel information. The location of the parcel map (cartographic data) in the jurisdiction is shown by reference to the coordinate system used in this GIS. The parcel numbers displayed on the map are the location identifiers that allow access into a descriptive data base of attribute data on each parcel (land use, dwelling units, and year built). In Figure 2.9, the cartographic data form a block map, the location identifiers are block numbers, and the attribute data consist of population characteristics.

Thus the references identified in Figures 2.8 and 2.9 (coordinates, cartographic data, and location identifiers) form the basis for using an urban geographic information system. When many departments and agencies (and different functions within those organizations) agree to share the use of a GIS, some of these references are common to all (such as streets or land parcels), while others are specific to the user (such as voting wards or sewer mains). Those references that are common are usually referred to as a *base map* because the different agency-specific geographic references are based upon the coordinates, map information, and location identifiers of these common references. Base maps of urban geographic information systems vary from one local government to another depending upon the needs of the users involved, but all base maps consist of three general references, which will be discussed in more detail in Chapter 6. These references include:

Coordinate systems, which relate map information to the earth. Examples include: state plane, universal transverse mercator (UTM), latitude and longitude, and locally defined systems that may apply only to the jurisdiction itself.

Cartographic data, which form the basic map information such as parcel maps maintained by a City or County Tax Assessor or Register of Deeds as well as right-of-way maps and certified surveys maintained by a City or County Engineer or Surveyor. Other cartographic data (blocks, streets, etc.) can be found in the centralized mapping functions of various city or county departments as well as public utilities, private companies, and state and federal agencies (the DIME File from the U.S. Bureau of the Census, for one).

Location identifiers, which uniquely identify map features across the entire jurisdiction. Unique parcel numbers, block numbers, census tracts, street segments, intersection codes, are examples.

The National Research Council (NRC), in its definition of the multipurpose cadastre (Panel on a Multipurpose Cadastre, 1980), however,

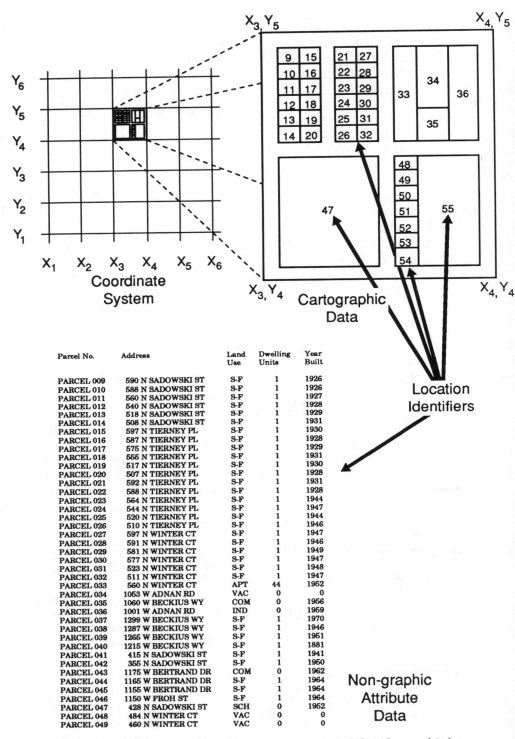

Parcel No.	Address	Land Use	Dwelling Units	Year Built
PARCEL 009	590 N SADOWSKI ST	S-F	1	1926
PARCEL 010	588 N SADOWSKI ST	S-F	1	1926
PARCEL 011	560 N SADOWSKI ST	S-F	1	1927
PARCEL 012	540 N SADOWSKI ST	S-F	1	1928
PARCEL 013	518 N SADOWSKI ST	S-F	1	1929
PARCEL 014	508 N SADOWSKI ST	S-F	1	1931
PARCEL 015	597 N TIERNEY PL	S-F	1	1930
PARCEL 016	587 N TIERNEY PL	S-F	1	1928
PARCEL 017	575 N TIERNEY PL	S-F	1	1929
PARCEL 018	555 N TIERNEY PL	S-F	1	1931
PARCEL 019	517 N TIERNEY PL	S-F	1	1930
PARCEL 020	507 N TIERNEY PL	S-F	1	1928
PARCEL 021	592 N TIERNEY PL	S-F	1	1931
PARCEL 022	588 N TIERNEY PL	S-F	1	1928
PARCEL 023	564 N TIERNEY PL	S-F	1	1944
PARCEL 024	544 N TIERNEY PL	S-F	1	1947
PARCEL 025	520 N TIERNEY PL	S-F	1	1944
PARCEL 026	510 N TIERNEY PL	S-F	1	1946
PARCEL 027	597 N WINTER CT	S-F	1	1947
PARCEL 028	591 N WINTER CT	S-F	1	1946
PARCEL 029	581 N WINTER CT	S-F	1	1949
PARCEL 030	577 N WINTER CT	S-F	1	1947
PARCEL 031	523 N WINTER CT	S-F	1	1948
PARCEL 032	511 N WINTER CT	S-F	1	1947
PARCEL 033	560 N WINTER CT	APT	44	1952
PARCEL 034	1053 W ADNAN RD	VAC	0	0
PARCEL 035	1060 W BECKIUS WY	COM	0	1956
PARCEL 036	1001 W ADNAN RD	IND	0	1959
PARCEL 037	1299 W BECKIUS WY	S-F	1	1970
PARCEL 038	1287 W BECKIUS WY	S-F	1	1946
PARCEL 039	1265 W BECKIUS WY	S-F	1	1951
PARCEL 040	1215 W BECKIUS WY	S-F	1	1881
PARCEL 041	415 N SADOWSKI ST	S-F	1	1941
PARCEL 042	355 N SADOWSKI ST	S-F	1	1950
PARCEL 043	1175 W BERTRAND DR	COM	0	1962
PARCEL 044	1165 W BERTRAND DR	S-F	1	1964
PARCEL 045	1155 W BERTRAND DR	S-F	1	1964
PARCEL 046	1150 W FROH ST	S-F	1	1964
PARCEL 047	428 N SADOWSKI ST	SCH	0	1952
PARCEL 048	484 N WINTER CT	VAC	0	0
PARCEL 049	460 N WINTER CT	VAC	0	0

FIGURE 2.8. Land records information for an urban GIS with parcel information.

views the base map differently. As described in Chapter 6, the multi-purpose cadastre proposed by the NRC consists of five essential components, one of which is "a series of current, accurate large-scale maps" (p. 14) often referred to as a base map that can be a "conventional photogrammetric line map" or "orthophoto map" (p. 53). Both types of maps are based upon the physical features represented by carto-

FIGURE 2.9. Land records information for an urban GIS with block information.

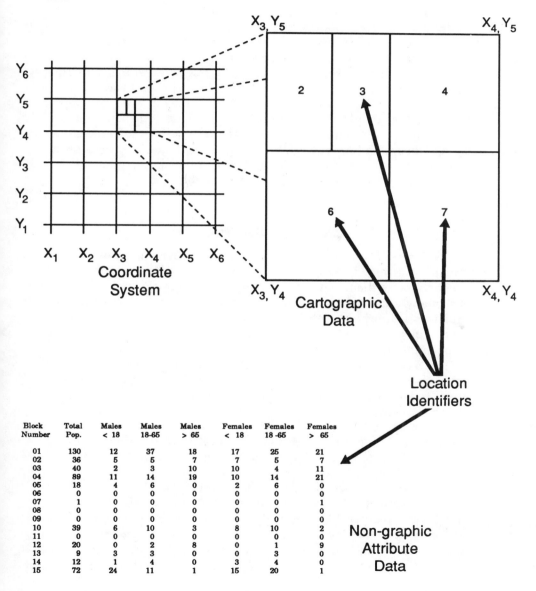

Block Number	Total Pop.	Males < 18	Males 18-65	Males > 65	Females < 18	Females 18-65	Females > 65
01	130	12	37	18	17	25	21
02	36	5	5	7	7	5	7
03	40	2	3	10	10	4	11
04	89	11	14	19	10	14	21
05	18	4	6	0	2	6	0
06	0	0	0	0	0	0	0
07	1	0	0	0	0	0	1
08	0	0	0	0	0	0	0
09	0	0	0	0	0	0	0
10	39	6	10	3	8	10	2
11	0	0	0	0	0	0	0
12	20	0	2	8	0	1	9
13	9	3	3	0	0	3	0
14	12	1	4	0	3	4	0
15	72	24	11	1	15	20	1

graphic data (streets, roads, highways, waterways, etc.) and, according to the NRC, must be "grid oriented, tied to the geodetic control system, and updated regularly." A base map, then, as defined by the NRC, consists solely of the cartographic data representing the physical features on the land, as long as it is associated with a coordinate system that relates map information to the earth and is also associated with parcel maps, parcel numbers, and related land data files. This focus separates the coordinate system and location identifiers from the concept of a base map and generally represents the base map within the context of a land information system, for the purpose of modernizing land records within local government.

Many geographic information systems in local government, however, do not have a focus on parcel maps, but do have base maps that are shared by many different users. Such maps can consist of street centerlines, address ranges, block outlines, physical features, and other data accessed universally by all users of the system. In this context, the base map of a GIS, whatever its uses are, consists of any cartographic data (whether they represent physical features or not) that are agreed to be of common value to all users—as long as it is referenced to a coordinate system that relates the cartographic data to locations on the earth and can be associated with location identifiers that uniquely identify the features represented by the cartographic data.

Topological Data Structures

The computer does not "see" the cartographic data as a human can see a map. It cannot look at a map of two intersections and see which street connects them as we can. We see a visual image and then know *implicitly* which street is common to both intersections. We learned these "spatial relationships" (left, right, next to, at, from, to, etc.) by the time we completed the first grade. Since then, we have been able to look at maps and know where features are in relation to other features.

The computer, on the other hand, needs to have these spatial relationships defined *explicitly* in a manner it can understand. To make up for its lack of cognitive skills, the computer depends upon explicit definitions of how features are related to each other before it can process data geographically.

For example, an automated mapping system can reproduce a map, and, to some extent, manipulate a shape on that map. But how does it know whether that shape is a polygon—a single geometric entity—or merely a set of lines that meet at their endpoints? How does it know that a long street is the boundary of three aldermanic districts? By defining these relationships explicitly. If the shape is defined as a polygon, then it knows its boundaries and can perform analyses on that basis (such as calculating the area or shading the inside). If, on the other hand, the shape is only a set of lines that meet, then the area of the shape cannot be automatically calculated nor shaded because the computer does not know what or where its boundaries are (see Figure 2.10). In order to perform polygon-related functions, then, the boundaries must be identified by the user at the time the function is needed. But that makes the system difficult to use. By establishing the definition of topological relationships (such as polygon boundaries, lines connected to a network, etc.) early in the data-base development process, applications using these relationships can be used much more

FIGURE 2.10. A polygon defined by a GIS without topological data structures [a] and with topological data structures [b].

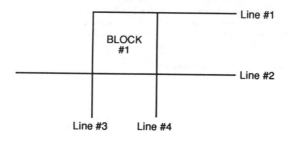

(a) Without Topological Data Structures

Block #1 is formed when Line #1, Line #2, Line #3, and Line #4 are displayed. It is not defined explicitly as a polygon - it only looks like a polygon <u>implicitly</u> to the human eye. The text "BLOCK #1" is displayed at a point which is defined when the map is digitized. Lines #1-#4 are not associated with Block #1.

(b) With Topological Data Structures

Block #1 is defined <u>explicitly</u> as a polygon whose boundaries are the line segments: Line #1, Line #3, Line #4, and Line #6. It is defined explicitly in a polygon table:

Polygon Name	Boundaries
Block #1	Line #1, Line #3, Line #4, Line #6

efficiently. The relationships are often defined to the system in the form of tables. The computer can "see" the values in these tables to understand the spatial relationships that the human can see on the map.

The rule, then, for designing geographic information systems is that anything of interest on a map must be *explicitly* defined as either a point, a line, or a polygon in order for the system to perform spatial analyses using the cartographic data (see Figure 2.11). (Text, annotations, and symbols are also usually defined explicitly either through

FIGURE 2.11. Geographic elements (points, lines, and polygons) used for defining topological data structures.

POINTS

are zero-dimensional objects on a map which represent a single location on the earth. The location is recorded as an xy coordinate. Depending upon the scale and accuracy desired, a point can represent the location of a multi-dimensional feature (such as a bridge).

Examples: street intersections
water valves
parcel centroids
addresses
fire hydrants

LINES

are one-dimensional objects on a map which represent a linear feature having a beginning point and an ending point. Depending upon the scale and accuracy desired, lines may be subdivided onto smaller units called arcs. (This is necessary in order to represent curved lines).

Examples: street centerlines
lot lines
rivers
water mains
sewer mains

POLYGONS

are two-dimensional objects on a map which represent shapes which have area. When the lines which form their boundaries are defined as polygon boundaries, the polygons become distinct objects which can be manipulated and displayed as single entities.

Examples: blocks
census tracts
parcels
zoning districts
political districts

the attribute data or as special tables defining their meanings and locations.)

Seeing these objects (points, lines, and polygons) on a map implicitly
is easy. Representing them for a geographic information system to use
requires explicit definition in the form of topology identified as zero-,
one-, and two-dimensional objects with the lower-dimensional objects
being used to define the higher-dimensional objects (Moellering, 1988,
p. 24). A general schema for defining the relationships is as follows:

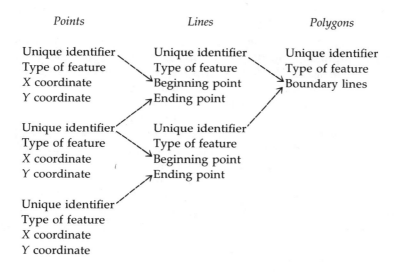

Points	Lines	Polygons
Unique identifier	Unique identifier	Unique identifier
Type of feature	Type of feature	Type of feature
X coordinate	Beginning point	Boundary lines
Y coordinate	Ending point	

Spatial Analysis Capabilities

Analyzing spatial data involves the determination of patterns of data
associated with locations and the manipulation of location-related data
to derive new information from existing data. Unwin (1983) describes
this as being concerned with spatial patterns defining the locational
relationships among points, lines, polygons, and surfaces and spatial
processes that define the dynamic nature of these features in terms of
distance, direction, and connectedness. Thus, the analysis of clusters
of tax delinquent properties, crime incidents, health problems, or water-
main breaks involves the determination of patterns that indicate non-
random occurrences and therefore require investigation of other
relationships to understand the reasons for the clustering. Spatial processes, on the other hand, involve the manipulation of data topologically (overlaying zoning boundaries onto parcel boundaries, determining the shortest path between a fire and a fire station, locating the
nearest valves to a water-main break, etc.) to create new information
from the relationships among the data and the topological structures
related to the data. In either case, spatial analysis begins with a prob-

lem or a question an analyst must address and then continues through an investigation of the map and attribute data that are available using various statistical, mathematical, geometric, and cartographic methods that can be referred to as *spatial analysis tools*.

These tools have been used for many years by geographers, cartographers, and others who must analyze information displayed on maps (Nystuen, 1968; Tufte, 1985; and Unwin, 1983). When geographic information systems technology made these tools available on computers, however, they became easier to use and allowed analysts to work interactively with the map and the attribute data in their investigations of problems and issues (Burrough, 1986).

The spatial analysis tools in a GIS are computer programs that retrieve, manipulate, and display information geographically. They normally involve access to the descriptive data bases for processing attribute data in addition to the mathematical calculations necessary to process the cartographic data base geometrically (coordinates, distances, and angles), utilizing the topological relationships of the points, lines, and polygons. Many geometric functions, such as map digitizing and updating, scale conversion, edge-matching, etc., are characteristic of automated mapping systems and should not be considered as spatial analysis tools. Similarly, attribute processing of nongraphic data, such as data entry, data-base retrieval, data updating, sorting, etc., are normally functions of data-base management systems and therefore are not unique to geographic information systems. It is the spatial manipulation of cartographic data together with attribute data that makes geographic information systems unique.

While these spatial manipulations of data can be used for many different analyses (Goodchild, 1978; Berry, 1987a, b; and Tomlin, 1983), they all begin with two rather basic questions identified by Nyerges and Dueker (1988):

1. Where is . . . (an object)?; and
2. What is at . . . (a location)?

The first question, looking for the location of an attribute value on a map, is simple without a computer—as long as the attribute value is displayed on the map. We have all looked at a map to locate a particular physical entity (a school, a park, a golf course, City Hall, etc.). This is because many maps have these entities and their descriptions plotted on them together with the cartographic information. But when the data are not displayed on the map, it is not possible to identify such entities visually and determine their location. A GIS, however, can find those attribute values in a data base and then relate them to the cartographic features on the map. Once they are found and related to the geography, then both the descriptive attribute values and the cartographic data can be manipulated by the spatial analysis programs of the GIS to identify other relationships and patterns.

The second question, looking for attributes of entities at a particular location (point, line, or polygon), is not so easily answered by looking at a map—unless that map has all the information about every possible entity of interest displayed on it. At the minimum, many different maps for the same geographic area, each with its own records (roads, water mains, crime statistics, etc.), would have to be researched and combined in order to answer this question. A GIS, however, can have multiple data bases of different types of information, all linked to the same geographic references, so that an inventory of information about entities at a point, on a line, or in a polygon can be available for analysis.

The spatial analysis capabilities of a GIS that can provide answers to these two questions also enable further manipulations on the data, allowing an interaction between the analyst and the data for investigating the relationships affecting the issues and problems of interest. These analysis capabilities have been documented and classified by many notable authors (most recently, Berry, 1987a, b; Densham and Goodchild, 1989; and Goodchild and Brusegard, 1989); however, the capabilities most useful to local governments can be grouped in accordance with the types of entities associated with the three topological features represented in the urban geography: points, lines, and polygons.

Analysis of Data Related to Points

Spatial query—Not truly an analysis process, but nonetheless the most often used function of an urban geographic information system, spatial query programs allow the user to find and display a map as well as the attribute data describing features on the map (such as a parcel). Figure 3.6 provides an example of spatial query for an address.

Geocoding—Also referred to as *geoprocessing*, this function provides an automated means to add geographic identifiers to the attribute data of features at particular points. Thus the attribute data for an address or parcel can be expanded automatically to include the census tract, police district, aldermanic district, etc., codes assigned by the system. Chapter 3 includes a description of how property records can be automatically updated with new zoning codes when zoning boundaries are changed.

Proximal analysis—Another powerful tool for saving time is the process of automatically displaying symbols of point-related features having certain attribute values and then identifying other features closest to them.

- *Dot-density mapping* allows the automatic display of symbols such as dots at the points of features that have certain user-defined characteristics. Dot-density maps of property tax delinquencies, crimes, survey returns by type of response, etc. are common uses of urban geographic information systems for visually identifying clusters of like occurrences. (See examples in Figures 3.13, 3.21, 3.22, 3.26, 3.27, 3.31, and 3.34.)

- *Nearest-neighbor* capabilities allow the identification of a feature that is closest to another feature. Thus the closest fire hydrant to a given building, for example, can be identified and highlighted.

Analysis of Data Related to Lines

Network analysis—The topological data structures of lines provide the capability to "follow" lines along an interconnected network and then process attribute data associated with those lines along the network. Water pressure districts, for example, are networks of individual water mains connected together. Network analysis programs are used to analyze data associated with the lines (water mains) such as water pressure. Similarly, *flow analysis* programs are used on sewer networks because they analyze attribute data (flow rate) of each sewer main in the network. *Routing* a vehicle along a network of streets, such as a fire truck going to a fire, is accomplished by identifying the street segments (lines) that connect the starting point (parcel number of the fire house) to the destination (parcel number of the fire location). When more than one route is possible, *optimum path analysis* programs use the attributes of the street segment lines (travel time, distance, etc.) to compute totals along each path and select the shortest distance or travel time between the two points (optimum path) based upon those totals.

Analysis of Data Related to Polygons

Polygon processing—A general term used to describe a number of capabilities using polygon boundaries to identify which point, line, or other polygon data to process. The polygon boundaries define the limits of the coordinates of points, lines, and polygons to process; those outside the boundaries are not processed, and those inside the boundaries are.

- *Point-in-polygon* functions identify which points are located within a polygon. The attributes of the features represented by those points can then be used to produce statistics, display symbols on a map, or print lists on a tabular report. (See examples in Figures 3.8, 3.9, and 3.16, and Tables 3.3 and 3.4.)
- *Choroplethic mapping* functions utilize the statistics computed from point-in-polygon functions (or computed or obtained from other sources) to shade or color the area within the boundaries of the polygon (also known as *polygon fill* functions). The shading pattern or color is determined by the magnitude of the value of the computed statistic (usually defined by a range of values). (See examples in Figures 3.28, 3.29, 3.30, 3.32, 3.33, 3.35, 3.36, and 3.37.)

- *Polygon overlay* functions use Boolean logic (AND, OR, NOT, etc.) to create new polygons from the boundaries of two or more different sets of polygons. The areas within the new polygons, then, have attributes that are Boolean combinations of the original sets of polygons. (See example in Figure 3.5.)

Polygonization—A term applied to the creation of polygons from data as opposed to the creation of polygon boundaries by digitizing boundaries. These polygons are created because of certain spatial criteria defined by the user.

- *Spatial aggregation* functions locate polygon boundaries from the results of spatial clustering such as shown on dot density maps. (Examples are provided in Figures 3.10, 3.11, 3.15, 3.17, 3.18, and 3.19.)
- *Buffering* is a function that forms a polygon around a point, line, or other polygon by placing its boundaries at a certain distance or other quantitative measurement from the feature (see example in Figure 3.38).

Exercises

1. Explain the differences between AM/FM, CADD, and GIS.

2. What are the tools available in GIS technology, and what are they used for?

3. Explain the difference between *implicit* and *explicit* spatial relationships and what they mean to a GIS.

4. Explain the difference between attribute data and cartographic data. Give examples.

5. How are data associated with geography in a GIS?

6. What makes a relational data base different from one that is hierarchical or network? Why would this be important in a GIS?

7. One use of an operating GIS is to display the total population for each city block on a map. When it is displayed, the block population total is overlaid onto the block number, rendering the data indiscernible. Why?

References

Berry, J.K. (1987a), "Computer-Assisted Map Analysis: Potentials and Pitfalls," *Photogrammetric Engineering and Remote Sensing*, Vol. LIII (10) (pp. 1405–10).
—— (1987b), "Fundamental Operations in Computer-Assisted Map Analysis," *International Journal of GIS*, 1 (pp. 119–36).
Burrough, P.A. (1986), *Principles of Geographical Information Systems for Land Re-*

sources Assessment (Monograph on Soil and Resources Survey, No. 12), Oxford University Press, Oxford, England.

Cowen, David J. (1987), "GIS vs. CAD vs. DBMS: What are the Differences?," *GIS '87 Proceedings Vol. I,* San Francisco, Calif. (pp. 10–18).

Dangermond, Jack (1988), "A Technical Architecture for GIS," *GIS/LIS '88 Proceedings—Volume 2,* ACSM, ASPRS, AAG, URISA (pp. 561–70).

Densham, Paul J., and Goodchild, Michael F. (1989), "Spatial Decision Support Systems: A Research Agenda," *GIS/LIS '89 Proceedings—Volume 2,* ACSM, ASPRS, AAG, URISA, and AM/FM International (pp. 707–16).

Exler, Ronald D. (1988), "Integrated Solutions for GIS/LIS Data Management," *GIS/LIS '88 Proceedings—Volume 2,* ACSM, ASPRS, AAG, URISA (pp. 814–24).

Goodchild, M.F. (1978), "Statistical Aspects of the Polygon Overlay Problem," *Harvard Papers on Geographic Information Systems* (ed. G. Dutton), Vol. 6, Addison-Wesley, Reading, Mass.

Goodchild, M.F. (1985), "Geographic Information Systems in Undergraduate Geography: A Contemporary Dilemma," *The Operational Geographer,* Vol. 8 (pp. 34–38).

Goodchild, M.F., and Brusegard, David (1989), "Spatial Analysis Using GIS: Seminar Workbook," National Center for Geographic Information and Analysis, AM/FM Conference XII, New Orleans.

Klein, Dennis H. (1988), "User Expectations for Selecting a Microcomputer-Based Municipal Automated Mapping/Land Information System (AM/LIS)," *GIS/LIS '88 Proceedings—Volume 2,* ACSM, ASPRS, AAG, URISA (pp. 550–60).

Martin, James (1976), *Principles of Data-Base Management Methodologies,* Prentice-Hall, Englewood Cliffs, N.J.

Martin, James (1977), *Computer Data-Base Organization,* 2e, Prentice-Hall, Englewood Cliffs, N.J.

Moellering, Harold (ed.) (1988), *Issues in Digital Cartographic Data Standards—Report #7,* National Committee for Digital Cartographic Data Standards, The Ohio State University, Columbus, Ohio.

National Science Foundation (1987), "The National Center for Geographic Information and Analysis: A Prospectus (Geography and Regional Science Program)."

Nyerges, Timothy L., and Dueker, Kenneth (1988), "Geographic Information Systems in Transportation," U.S. Department of Transportation, Washington, D.C.

Nystuen, J.D. (1968), "Identification of Some Fundamental Practical Concepts," *Spatial Analysis—Reader in Statistical Geography* (Berry, J.L.B., and Marble, D.F., eds.), Prentice Hall, Englewood Cliffs, N.J.

Panel on a Multipurpose Cadastre, Committee on Geodesy, Assembly of Mathematical and Physical Sciences, National Research Council (1980), *Need for a Multipurpose Cadastre,* National Academy Press, Washington, D.C.

Tomlin, C.D. (1983), "A Map Algebra," *Harvard Computer Graphics Conference Proceedings,* Harvard University Graduate School of Design, Cambridge, Mass.

Tufte, E.R. (1985), *The Visual Display of Quantitative Information,* Graphics Press, Cheshire, Conn.

Unwin, D. (1983), *Introductory Spatial Analysis,* Methuen, London.

ADDITIONAL READINGS

Dahlberg, R.E., McLaughlin, J.D., Niemann, B.J., Jr. (1989), *Developments in Land Information Management*, Institute for Land Information, Washington, D.C.

Dangermond, Jack, and Freedman, Carol (1984), "Findings Regarding a Conceptual Model of a Municipal Data Base and Implications for Software Design," *Modernizing Land Information*, seminar on the multipurpose cadastre, Institute for Environmental Studies Report 123, University of Wisconsin-Madison, Madison (pp. 12–49).

Dangermond, Jack, and Freedman, Carol (1987), "Introduction to Geographic Information System Technology," *NCGA'S Mapping and Geographic Information Systems '87 Proceedings*, National Computer Graphics Association, Washington, D.C. (pp. 32–44).

3

Applications of Urban Geographic
Information Systems

The key to making computers useful to management is learning
how to present the right information in the right way, and that is
no simple matter.

MARTIN, p. 292

One plot is worth 1000 pages of printout.

UNKNOWN

The single most distinguishing characteristic of an urban geographic
information system is its ability to integrate information from many
different sources and at many different levels of responsibility in an
organization. Geography itself is a common reference used by virtually
every activity in local government, whether it be to find a water main
valve, set property tax assessments, or begin a new solid waste recycling program. Maps and data associated with locations are resources
used daily for delivering public services, managing public resources,
and setting public policy in all local governments across the nation and
throughout the world. Geographic information systems are used by
many of these local governments to improve the services they provide
and the decisions they make for the public good.

One local government that has been using an urban geographic information system for many years is the City of Milwaukee. Installed in
1976, the Milwaukee system has been used in a wide range of applications in the service-delivery, management, and policy-setting functions of the city. This chapter provides a detailed discussion of some
of these applications, ranging from the daily activities of recording
changes to maps, to the annual review of liquor licenses, to the decennial reapportionment after a census, to the once-in-a-lifetime switch
from garbage cans to garbage carts. They range from service delivery
(water-main breaks) to policy setting (lead-based paint removal); individual property records (property ownership maps) to city-wide trends
(housing quality); from data retrieval (building permit review) to data
analysis (neighborhood library siting). The figures included with these
descriptions are reprints from actual products used by public officials

and thus represent real data used in an operational setting of urban government.

Many of the functions described in this chapter have been performed by local governments throughout the nation for 100 years or more. Performing them with the aid of a GIS, however, has helped make Milwaukee more efficient, more responsive, and more effective in serving its citizens than those governments still struggling with manual, segregated, and outdated records.

Map Updating

Plat Maps

A deed contains a description of a parcel of land that accurately and precisely locates the parcel. As described in Chapter 6, the deed defines the property boundaries in relation to other legal references (such as lot and block of a subdivision or distance and bearing from a survey monument). Land surveyors utilize these descriptions to prepare land subdivision plats, assessor's plats, cemetery plats, and certified survey maps that are legally recorded in files of the County Surveyor. These graphic representations of the legal property descriptions are used by local government for a number of purposes, including the assessment of properties for taxing their owners. Figure 3.1 is an example of a typical assessor's plat map of one block in an urban setting that was created from the legal descriptions of the deeds for the properties in this block.

The block is bounded by four streets: W. Villard Ave., N. 68th St., W. Lancaster Ave., and N. 69th St. It is Block 5 of the Jefferson Gardens subdivision. The boundaries of each property in the block are drawn on this map, and each property is identified in two manners: the lot number (Lot 1 through Lot 34) and the parcel number (the parcel number of Lot 1 is 212-0197). The parcel number identifies the lot or combination of lots that are under one ownership. Notice that Lot 5 and Lot 6 are owned by the same person and have been "combined" into Parcel Number 212-0201. Notice, also, that Lot 18 has been "split" and combined with its two adjacent lots (17 and 19), forming two ownership parcels (Parcel Numbers 212-0212 and 212-0213). The "vinculum" is a symbol used to show that a lot has been combined with other lots or portions of other lots. The dashed line is used to show that a parcel has been split into two ownerships.

The lot number, together with the block number and subdivision name, form the "legal description" of a property and are recorded in the text of the deed. The plat map is only a graphical representation of this legal description—the deed contains the official record.

Changes to plat maps occur when:

SCALE 1 IN.=80 FT.

JEFFERSON GARDENS
VOL. 44   P. 77

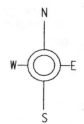

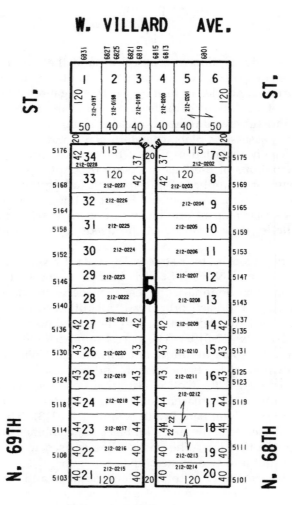

FIGURE 3.1. An assessor's plat map.

1. The legal description of a deed changes the size of the property (a change in the owner name, however, does not change the legal description);
2. A new subdivision is created (even under one ownership) because new lots are created;
3. A change is made to the public right-of-way (new streets, new alleys, widened streets or alleys, vacated streets and alleys that are converted back to private ownership, etc.) because they cause changes to the lot and block lines;
4. Annexations cause new plat maps to be created;
5. Other changes to the information on the map occur that do not relate to the boundaries of the properties:
   a. Changes in street names;
   b. Changes in addresses;
   c. Corrections to errors on the map.

The volume of these changes varies from jurisdiction to jurisdiction and depends upon: the size of the jurisdiction, the amount of growth and land development it is experiencing, and the amount of redevelopment in older areas. Plat maps are usually created and maintained by the Tax Assessor, Register of Deeds, or Property Lister in local governments.

Quarter-Section Maps

Another series of maps widely used in local government is the "quarter-section" map series. Quarter-section maps are ½ mile by ½ mile square graphical representation of legal surveys of the land. The information recorded on these maps include: all public right-of-way (including "vacated" and "dedicated" streets and alleys that are not physical streets and alleys but that were at one time or will be at a future date); all approved subdivisions, certified survey maps, and land divisions; other geographic references such as street names, rivers and shorelines, and public corporation limit boundaries; and text annotations that assist in interpreting the map. Figure 3.2 is an example of a quarter-section map. Notice that subdivision information on this map is extensive, including the subdivision boundaries and dimensions, the subdivision name, block numbers, lot numbers, and lot dimensions. Certified survey map information includes CSM number, boundaries, and dimensions.

Property ownership boundaries are not recorded on these maps as they are on plat maps. That is because quarter-section maps record the results of physical surveys of the land that are a legal definition of the units of the land and do not define ownership of those units.

Changes to quarter-section maps occur when:

1. A new subdivision is created in preparation for new development (or redevelopment);

FIGURE 3.2. A quarter-section map.

2. A new survey of an existing subdivision is conducted to change lot boundaries or correct errors in a previous survey;
3. A change is made in the size or location of the public right-of-way (new streets and alleys or changes to either);
4. A street or alley is vacated (no longer open for public use);
5. A street name is changed;
6. Property outside the jurisdiction is annexed;
7. Errors of any kind are detected and corrected.

As with plat maps, the volume of changes to these maps depends upon the size, growth, and amount of redevelopment activities of the jurisdiction. Quarter-section maps are normally created and maintained by a City Engineer, County Engineer, or County Surveyor.

Land Use Maps

Yet another common series of maps used in local government are land use maps (see Figure 3.3). Land use maps record how the land in a jurisdiction is used by the citizens: whether it is residential, commercial, industrial, public, open, etc. While the map in Figure 3.3 shows symbols identifying these land use categories for each parcel and, specifically, each structure on the parcel, many local agencies map land use information on larger geographic areas such as blocks or portions of blocks. Except for the land use symbols themselves, most of the

FIGURE 3.3. A land use quarter-section map.

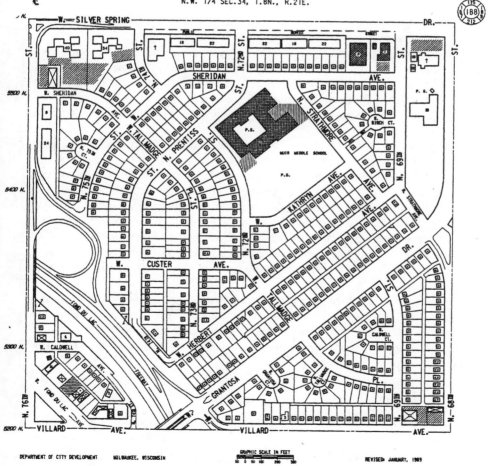

information on these maps is the same as the information recorded on plat maps and quarter-section maps: parcel ownership boundaries, public right-of-way, street names, and annotations.

Changes to land use maps occur when:

1. Buildings are built and demolished;
2. Existing buildings are renovated so as to change the way they are used (e.g., converting a house into a commercial enterprise without demolishing it and rebuilding);
3. Property ownership boundaries change;
4. New subdivisions are created;
5. A change is made in the size or location of the public right-of-way;
6. A street or alley is vacated;
7. A street name is changed;
8. Property outside the jurisdiction is annexed;
9. Errors are detected and corrected.

Unlike plat maps and quarter-section maps, the volume of changes to land use maps depends upon the physical changes to structures on the land. While all changes to plat maps and quarter-section maps also cause the land use maps to change (when the jurisdiction records land use by parcel as shown in Figure 3.3), changes to the structures on the parcels cause additional changes to the land use maps. Generally, the local Planning Department of the agency creates and maintains these maps.

Obviously, from the descriptions of these three map series—the plat maps, the quarter-section maps, and the land use maps—much of the map information recorded on them is redundant (street names, parcel boundaries, right-of-way). Since they are usually maintained in three separate functional units, this redundant information must also be maintained redundantly by the separate units. Thus, one change can cause three separate update actions to take place. Depending upon their own local needs, local governments maintain many more map series than these three, causing an even greater degree of redundant map updating.

A geographic information system eliminates most of this redundancy by storing all map information (regardless of the functional unit responsible for each map) in one data base where only one change need be made to common data. The change is then reflected on all maps containing that information.

Zoning

Most local governments use zoning regulations to guide growth and development within their jurisdictions in accordance with officially adopted long-range land use plans. While land use plans, developed by urban planners and approved by elected officials (and, in some cases, regional planning agencies), form the basis upon which future use of the land will be made, it is zoning, legally adopted restrictions on specific uses and building sizes in specific geographic areas, that allows public officials to approve or reject specific development proposals. The city's Code of Zoning Ordinances provides the legal descriptions of those restrictions and the locations where those restrictions apply. The zoning map represents a geographic display of the contents of the zoning ordinances.

Traditionally, the zoning map has been used as a record of changes to the Code of Zoning Ordinances *and* as a means to enforce the Code when physical changes to the city are proposed by developers and property owners. This enforcement of the Code is performed during the building permit review process. Both functions, *recording* the zoning restrictions and *enforcing* those restrictions, were an integral part of the design of Milwaukee's geographic information system.

Recording Zoning Changes

Proposed changes to the Code of Zoning Ordinances and the Zoning Map are generally made by developers who desire to build on land where existing zoning regulations prevent the type and size of buildings that the developer wishes to construct. They are also proposed by public officials who, for various reasons, feel that existing regulations are either too strict or too lenient for the interests of the neighborhood. Whatever the reason for proposing changes to existing zoning, each specific proposal is carefully researched and formally proposed to the city's Zoning and Development Committee, composed of representatives of the Common Council. Once the Zoning and Development Committee approves a proposed change to the Code of Zoning Ordinances, Common Council provides legal approval of the change by adopting a resolution that amends the Code of Zoning Ordinances.

The proposed change is presented in graphical form, displaying the existing land use and zoning of the area as well as the surrounding area, and the proposed zoning of the area (see Figure 3.4b). It is also presented in textual form, describing the legal description of the zoning boundaries for changing the ordinance (see Figure 3.4a). These legal descriptions are given in metes and bounds form, usually in reference to the centerline of a street or other feature that can be legally surveyed for location. It is the responsibility of the city's Planning Department to not only prepare the proposed change information and map, but also to change the official zoning map after the change is adopted by Common Council.

Unlike most communities that utilize one set of zoning districts maps, the City of Milwaukee maintained an archaic system of districting for zoning that utilized three types of zoning districts specified in the Code

COMMON COUNCIL
CITY OF MILWAUKEE

FILE NUMBER: 882209

The Mayor and Common Council of the City of Milwaukee do ordain as follows:

Part 1.   There is hereby added to the Milwaukee Code a new section to read as follows:

Section 295-320(8).0021.   The zoning map is hereby amended so as to change the zoning for the area bounded and described as follows:

Commencing at the point of intersection of the center line of West Meinecke Avenue and the center line of North 27th Street; thence Southerly along the center line of North 27th Street to a point 120 feet north of the north line of West North Avenue; thence Easterly and parallel to West North Avenue to the center line of North 26th Street; thence North along the center line of North 26th Street to a point 183.33 feet north of the north line of West North Avenue; thence easterly and parallel to West North Avenue to a point 122 feet east of the east line of North 26th Street; thence Southerly and parallel to North 26th Street to a point 120 feet north of the north line of West North Avenue; thence Easterly and parallel to West North Avenue to the center line of North 25th Street; thence Northerly along the center line of North 25th Street to a point 64 feet south of the south line of West Meinecke Avenue; thence Westerly and parallel to West Meinecke Avenue to a point on the west line of North 25th Street; thence Southwesterly to a point which is 29 feet west of the west line of North 25th Street and 78 feet south of the south line of West Meinecke Avenue; thence south and parallel to North 25th Street 26 feet to a point; thence Northwesterly to a point which is 73.40 feet south of the south line of West Meinecke Avenue and 122 feet west of the west line of North 25th Street; thence continuing Northwesterly along the previously described line to the center line of the north-south alley lying between North 25th Street and North 26th Street; thence Northerly along the center line of said alley to the center line of West Meinecke Avenue; thence Westerly along the center line of West Meinecke Avenue to the point of commencement from Multi-Family Residence (R/D/40) to Local Business (L/D/40).

(a)

of Ordinances: use, area, and height. Zoning *use* restricts the type of activity conducted on the property (residential, commercial, manufacturing, etc.). The primary purpose of zoning laws is to preserve neighborhood homogeneity and encourage growth in accordance with the city's long-range plan. Allowing a cement-producing plant to be built in a single-family residential neighborhood, for example, is not consis-

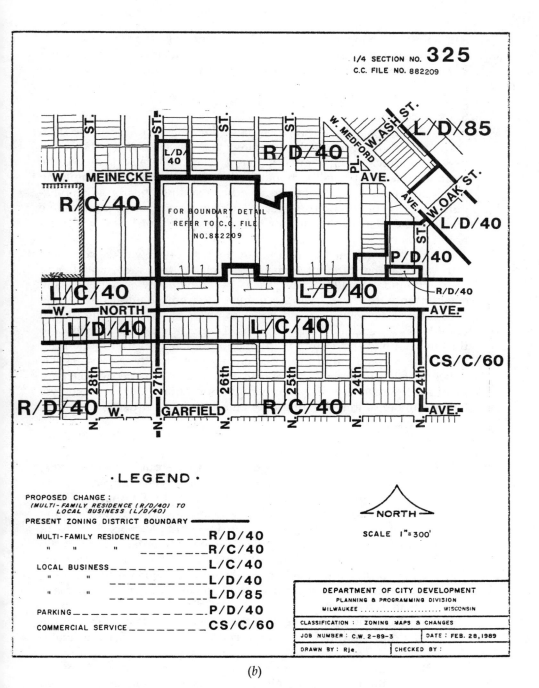

FIGURE 3.4. Proposed zoning change. (*a*) A metes and bounds description and (*b*) a graphical representation.

tent with preserving neighborhood homogeneity (nor property values). Similarly, preventing construction of commercial establishments along a major arterial highway (away from residential neighborhoods) prevents the economic growth of a jurisdiction. *Area* restrictions are imposed to preserve the aesthetic quality of neighborhoods by limiting the size of a building to an acceptable proportion of the size of parcel upon which it is built, setting standards for "set-back" (distance from the street to the front of the building) and "side-yard" (distance between buildings along a street). *Height* restrictions also preserve the aesthetic quality of neighborhoods by limiting the height of a building to be consistent with the height of other buildings nearby. Combinations of different zoning use, area, and height restrictions, then, define the physical plan of the city for balancing quality of living with economic vitality.

For many years, the zoning map at the City of Milwaukee consisted of a set of three different maps: one each for use, area, and height district boundaries. Physically, it was 14 feet wide by 50 feet long and covered an entire wall, using a unique electromechanical device that allowed it to be rolled up and down in order to locate zoning district boundaries for each of the three maps. There were two such devices in the city that had the zoning map: one in the Planning Department for use in the preparation of proposed changes and for the review of new development proposals, and one in the Building Inspection Department for the review of building permit applications. This arrangement had a number of longstanding problems associated with it:

1. Zoning district boundary changes had to be recorded six times (three for each map on both devices);
2. Changes to the base map information (the public right-of-way) had to be recorded independently from the changes in the Tax Department (tax plat page maps), the City Engineer's Office (quarter-section maps), and the land use section of the Planning Department (land use maps), causing multiple map changes in five different offices for a single change in the public right-of-way;
3. The physical and organizational separation of the two zoning maps made coordination of changes difficult and, sometimes, inconsistent because there was no method for ensuring that both maps were in agreement;
4. The electromechanical devices transporting the zoning map often broke down through wear and tear and were expensive to maintain;
5. The zoning map did not contain property lines nor addresses, causing the building permit review process for buildings located near zoning district boundaries to require additional time-consuming research.

The city's geographic information system offered an ideal solution to these problems. Each of the three zoning maps (use, area, and height) was used to digitize zoning district boundaries onto three separate levels of the digital base map, which already contained (along with other information) public right-of-way, property boundaries, parcel addresses, and parcel identifiers (tax key numbers) from previous map conversion projects. The boundaries of each type of zoning district were digitized as polygons and then overlaid onto each other (through polygon processing programs) to produce a single level with unique polygons for each combination of use, area, and height restriction. The electromechanical devices were discarded and replaced with GIS work stations for accessing and updating the zoning map information. Changes to the base map information (the other levels in the digital map data base) were recorded by other departments when they occurred and then automatically were available for display on the zoning map.

The benefits of incorporating the zoning map information in the city's GIS were profound. The time required to prepare proposed zoning changes (as shown in Figure 3.4) was reduced from 4 hours to only 1½ hours. Redundant map maintenance for approved zoning changes and for changes in the public right-of-way was eliminated. Inconsistencies in the zoning map information between the Planning Department and the Building Inspection Department were eliminated because zoning information resided in a single, centralized data base that was accessible by both departments. The cost of maintaining the two electromechanical devices (about $30,000 per year) was eliminated.

In addition to these benefits, the city's GIS made it possible, for the first time, to relate zoning boundaries to individual properties and lot lines that had been digitized from the tax plat maps (see Figure 3.5). Previously, the metes and bounds descriptions of zoning boundary lines, based upon the offset or distance along street centerlines, prevented the accurate placement of zoning boundaries in relation to individual lot lines. This meant that many properties—those that lay on the boundaries of zoning areas—actually were in two different zoning areas. (In some cases, a garage of a single-family home was located in a local business zone while the house was in a residential zone and the resident operated an automotive repair business out of his garage!) Because the tax plat maps and the zoning map were digitized on the city's GIS using the same coordinate system (state plane coordinates), individual property lines could be overlaid onto zoning boundaries to identify inconsistencies. In Figure 3.5, for example, an inconsistency exists between the zoning boundaries of the "R/B/60" district and the "L/B/60" district in the upper left-hand quadrant of the map. Notice that a large 80-unit apartment building (identified by an "80" located inside the building outline) is split between the two districts. Since the "R" in the R/B/60 district represents residential use and the "L" in the

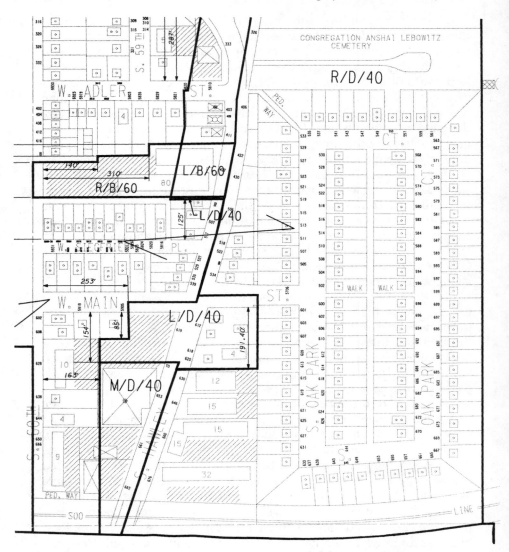

FIGURE 3.5. Zoning map showing nonconforming land use.

L/B/60 district represents local business use, the boundary of the L/B/ 60 district is wrong, albeit legally binding. Notice also that the small "L/D/40" sliver is meaningless because it is too small for any building to be built inside it. Obviously, these two inconsistencies show that the L/B/60 district should have the northern property line of the apartment building as its southern boundary in order for the zoning map and existing land use to be consistent. Other inconsistencies (nonconforming land uses) can also be identified, since the city's land use maps, showing land use by property, were also digitized on the system. Officials can identify individual properties whose existing land use does not conform to the zoning use prescribed for the area.

An additional benefit of having the zoning map on the city's GIS is that the nongraphics property records of the Tax Assessor's Office are automatically updated with new zoning codes when the zoning map is changed. Since property numbers (tax key numbers) are contained in the digital map base and since zoning districts are defined as polygons, all tax key numbers located within a zoning district that has been changed are identified by polygon processing programs. These tax key numbers and the new zoning code are then passed to the nongraphics data base to update the records with the new zoning code. This process eliminates the need for entering the new codes for each property on computer terminals.

Enforcing Zoning Restrictions

The city's Building Inspection Department enforces the Zoning Code of Ordinances through the building permit application process. Typically, a developer or contractor presents construction plans for changes to individual properties to the Building Inspection Department by completing a building permit application form. One of the first steps taken by the department in reviewing the permit application is to compare the proposed changes with the legally adopted zoning restrictions for the area in which the property is located. If the proposed change in use, building size, or building height conform to existing zoning regulations, the permit is approved and construction can begin.

Before the city had a GIS, the building permit review process was difficult because the zoning map prepared by the Planning Department did not have individual properties located on it. Since the information provided by the contractor or developer on the permit application form could identify the property only by address, finding that address on the zoning map, which did not have property boundaries, was difficult, especially when the property was located close to a zoning boundary. Usually, Building Inspection Department personnel had to confer with the Tax Assessor's office when this happened.

The city's GIS made this part of the permit review process easier in two ways. First, as described above, all changes to the zoning map could identify each parcel whose zoning was changed. This meant that, even without a map, the zoning of a property (accessible on the computer by address) could be identified. Second, since individual land use by property could be overlaid on the zoning map, and since individual maps could be accessed by address, the zoning *and* land use for each property could be displayed in map form. Figure 3.6 shows how the city's GIS is used to make the building permit application review process by address easier.

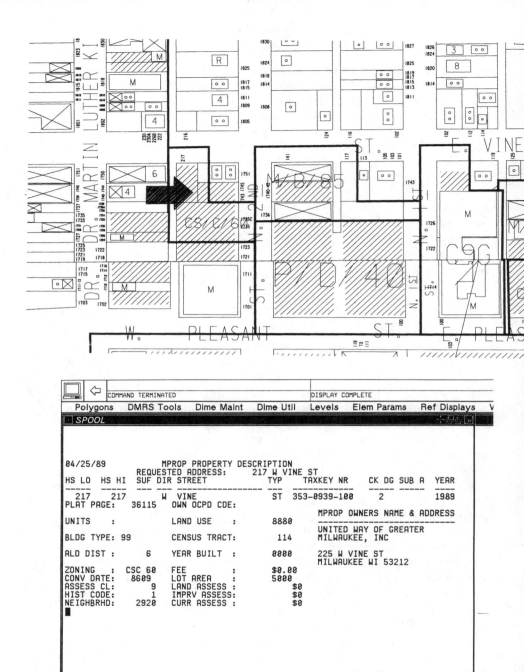

FIGURE 3.6. The building permit application review process requires both a zoning map and a nongraphics attribute display for a property.

Figure 3.6 represents the results of a building permit application inquiry to the city's GIS for a property at 217 W. Vine St. Not only are the property attribute data for that address displayed (including the zoning of CS/C/60), but the zoning map with land use for that property is also displayed with an arrow pointing to the property at that address.

Reapportionment

After each decennial census, the U.S. Bureau of the Census provides each state with new population figures, which are then used to redraw various federal, state, and local political districts. Because of significant population shifts that occur over the 10-year period between each census, there is usually a high level of interest at all levels of government in the impact that the population changes have on political districts. The impact of these changes is widespread: from the number of dollars funneled from the federal government to state government and from state government to local government; to changes in elected officials because their residences no longer lie within their district boundaries. No wonder public officials—elected and appointed—constantly keep tabs on the reapportionment process once the population data are provided by the Census Bureau.

In Milwaukee the reapportionment process is no different. First, the city's Election Commission must ensure balanced population by voting ward so that workload is balanced among polling places on voting day (long lines and voting delays always attract media attention!). In the process, local aldermen, attempting to preserve their loyal constituent base, monitor the movement of their district boundaries caused by changes in the voting ward boundaries (some even present their own plans for consideration). Once the city's changes in voting wards and aldermanic districts are formally adopted through Common Council resolution, the county takes the new boundaries and builds its supervisory district boundaries. Finally, the state completes the process by balancing population of supervisory districts among new state and federal representative and senatorial districts. Because of the widespread interest in these various boundary changes, the pressure begins to build on the city's Election Commission to complete their process soon after they receive the census data. The more rapidly this process goes, the less conflict there is among the interested officials.

Prior to 1980, the city used scores of clerical people to annotate population figures from the Census Bureau onto a large wall-sized map of the city that showed block outlines and existing voting ward boundaries. The population of each block, as provided by the census data, was hand-drawn onto this map (about 8000 blocks in all). Then the clerical people added the population totals for all the blocks within each voting ward (at best, with electronic calculators) to obtain the latest population totals for each voting ward. Where the totals were significantly different, the ward boundaries were moved (taking into consideration the locations of available polling places), and the population totals recalculated. Once the new voting ward boundaries were set, the tedious task of assigning them to aldermanic districts was begun (state law required voting wards to lie wholly within aldermanic districts and therefore could not be split between districts). Since this was a slow, manual process, it is not hard to imagine all 16 aldermen looking over the shoulders of the clerks as they drew lines and calculated population totals, with the aldermen suggesting changes that caused redrawing and recalculating. This process, in 1970, took 6 months and cost over $60,000 in clerical time.

Beginning with the 1980 Census, the city's GIS took control. After the 1980 enumeration process was complete, the Census Bureau provided a computer tape containing a flat file of population totals by block, giving a census block number and a population total for each of the approximately 8000 blocks in the city. Since one of the location identifiers on the city's GIS cartographic data base was census block number (see Figure 2.9), it was a simple procedure to match the block number on the tape to the block number on the cartographic data base so that the block population could be located and displayed on a map showing block outlines (see Figure 3.7).

No longer were clerks needed to draw the population figures in each block on the wall-sized map—the numbers were stored in the GIS at the locations of the blocks they represented. The polygon processing features of the GIS allowed the aggregation of block population figures to larger geographic areas (in this case, block groups). Figure 3.8 shows this aggregation process. These block group boundaries were then adjusted interactively by combining block groups or by moving their boundaries along block boundaries, obtaining new block group population totals each time a boundary change was made.

This process also required the display of potential polling places (schools, fire houses, and other public buildings) to ensure that there was a convenient location for the residents to vote. In order to identify these potential polling places, the city's Master Property File was accessed by land use code (school, fire house, etc.) to obtain their parcel numbers. The parcels with those land use codes were then displayed on the block group map at the location of their parcel numbers as an overlay to determine visually if they could be used as polling places. Election Commission personnel then visited these potential sites to determine which would be used.

The result of this process of balancing population and locating polling places is shown in Figure 3.9. The population within each voting ward is shown as the lower number in each pair, and the voting ward number is shown as the upper number. Each voting ward boundary is displayed in heavy dotted lines, and the location of each polling place is displayed as a box with the letter "P" inside.

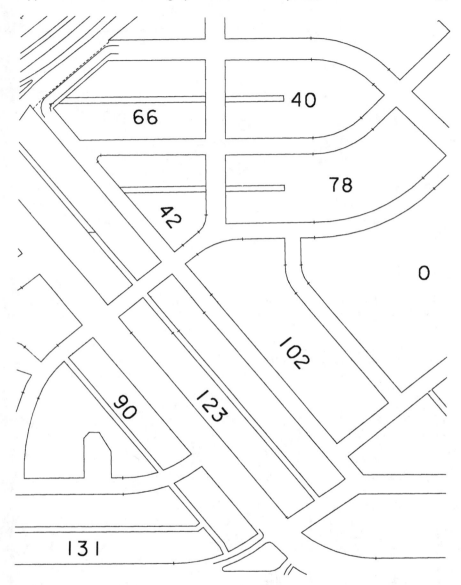

FIGURE 3.7. Display of block population totals for use in the reapportionment process.

Once voting ward boundaries were established, the process of establishing aldermanic district boundaries was similar because it consisted of using the population of each voting ward to balance population among the 16 aldermanic districts. Thus, instead of using the block populations to build larger geographic areas, the voting ward populations were

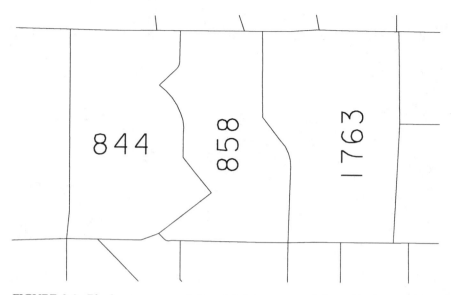

FIGURE 3.8. Block group population totals aggregated from block population totals.

FIGURE 3.9. Voting wards with balanced population totals built from block groups showing total population, voting ward numbers, and locations of polling places [P].

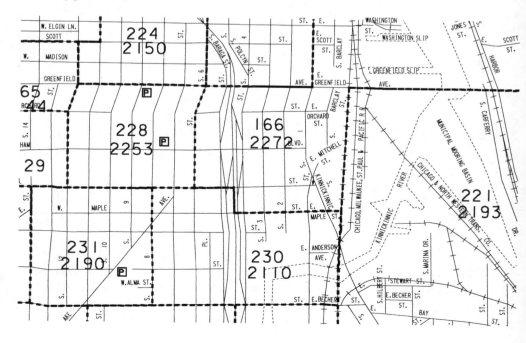

used to build aldermanic districts. Not only did state law require voting wards to lie wholly within aldermanic districts, but it also required that the aldermanic district populations deviate no more than 3 percent. This required an interactive process of digitizing "potential" aldermanic district boundaries along voting ward boundaries to complete a polygon, allowing the polygon processing programs to compute population totals for the districts. Where the figures deviated by more than 3 percent, the aldermanic district boundaries were moved to different voting ward boundary lines and population totals recalculated—all done interactively at the GIS work station.

The result of this process is shown in Figure 3.10: new aldermanic districts that vary no more than 3 percent in population and whose boundaries are coterminous with voting ward boundaries. Figure 3.11 is the official Election Commission map for Aldermanic District Number 8, showing the district boundaries, all streets and address ranges, voting ward numbers and boundaries, and polling places for the district. These maps are used for public display to help city residents identify to which voting ward they are assigned and where they should vote.

The cost of this process? It took 3 months and $24,000 in 1980 to complete the process and produce the 16 aldermanic district maps as shown in Figure 3.11. This was a significant improvement over the 6 months and $60,000 required after the 1970 Census.

Building Inspection Workload Balancing

The Building Inspection Department, among other responsibilities, inspects properties to ensure that their electrical work is safe in accordance with established building codes. The properties to be inspected are identified through the building permit process. In one year, a large city could conduct over 30,000 electrical inspections due to permit applications. With such a sizeable amount of work, the electrical inspectors are assigned to geographical subunits of the city (electrical inspection districts), and in the City of Milwaukee, each inspector is assigned to one district—13 in all (see Figure 3.12). Since over the years the number and size of buildings within each district change (because of demolitions, new construction, renovations, etc.), the number of inspections to be conducted also changes, causing an imbalance of workload among inspectors. One inspector, it was found, had three times as many inspections to conduct as another. In order to balance the workload, the district boundaries are periodically changed, based upon the number of permits in each district.

Without a GIS, and using manual methods, adjusting inspection district boundaries in order to balance workload required one full week with all 13 inspectors together in a room, trading cards (one for each permit) until there were 13 even stacks and each stack defined one

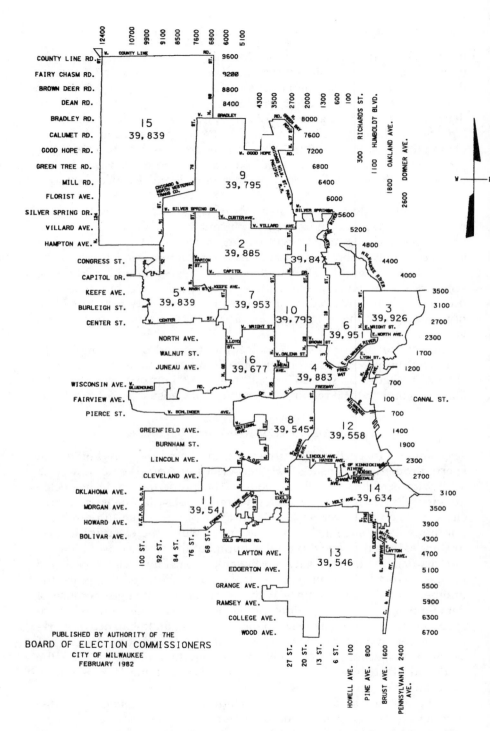

FIGURE 3.10. New aldermanic districts with balanced population totals.

84

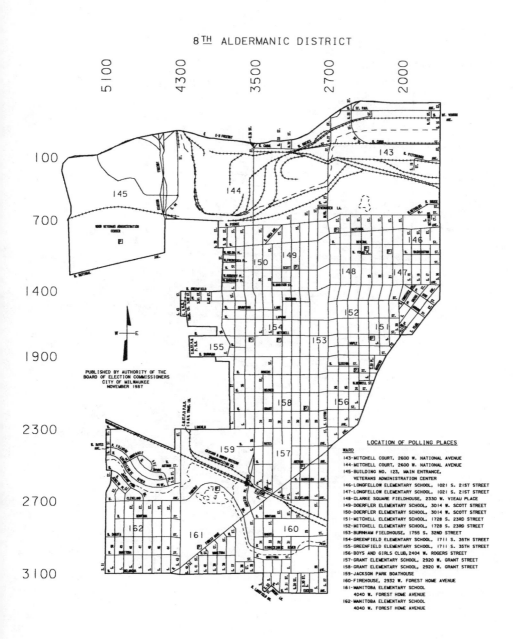

FIGURE 3.11. Election commission map of Aldermanic District 8.

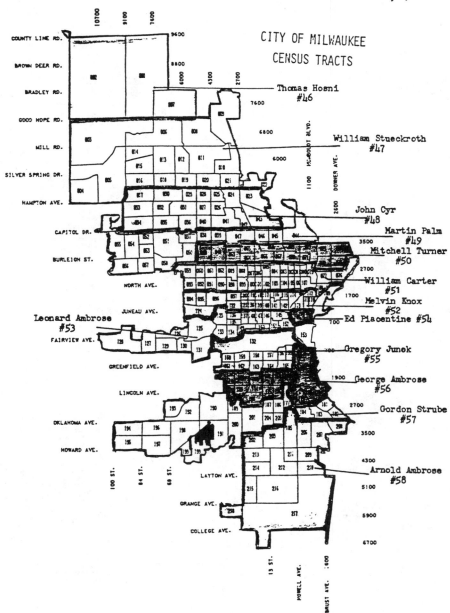

FIGURE 3.12. Old building inspection districts. Changes in construction and demolition activity caused imbalances in the workload of inspectors.

86

contiguous geographic area. Since this process was very labor intensive (480 work hours) and took the inspectors away from their primary responsibilities, it was not conducted very often. As a result, inspector workload was usually out of balance, causing morale problems in the department.

One year, departmental personnel decided to try a different method for redefining the inspection district boundaries to balance the workload among the 13 inspectors—one that would not require all 13 to meet together for a week. Their supervisor obtained a map of the city with census tract boundaries displayed and a special computer listing summarizing permit data relating to electrical inspections by census tract. Assuming that new districts could be determined by combining census tracts, the supervisor began to add the statistics for groups of census tracts together so that the groups had roughly the same number of inspections. By working with the maps and the summary data, the supervisor could redistrict by himself and not perform the time-consuming physical process of stacking cards. He gave up after 3½ hours of writing numbers on the map, drawing temporary boundaries, adding the numbers, changing the temporary boundaries, subtracting, and adding the numbers again. The process took too long and still would not accurately balance the workload because census tracts were too large. Moving a census tract from one district to another caused too large a change in the district totals, and the numbers could not be broken down into smaller geographic units.

Fortunately, the city had a geographic information system with "redistricting" capabilities designed specifically for this type of application. The computerized building permit records of the Building Inspection Department were accessed to obtain the addresses of all permits requiring electrical inspections over the past two years. (Two years of data were needed to project future work adequately because construction activity does not radically change in such a short time period.) These addresses were then matched to an index file that provided location identifiers (parcel numbers), allowing the permits to be linked automatically to the city's geographic data base. This allowed the GIS to display the spatial distribution of the permits on a map of the city, as shown in Figure 3.13.

While each dot on the map in Figure 3.13 represents a unique parcel (and address), it may not represent a unique permit since some building had more than one permit (think of a large office building or apartment building with a number of different units). To represent workload accurately, then, the redistricting software of the GIS counts the number of permits for each unique parcel number and displays that number on the map. Thus, each dot is actually the total number of permits issued for a parcel (the scale of the map in Figure 3.13 is so small that the number appears visually as a symbol—a dot). To obtain

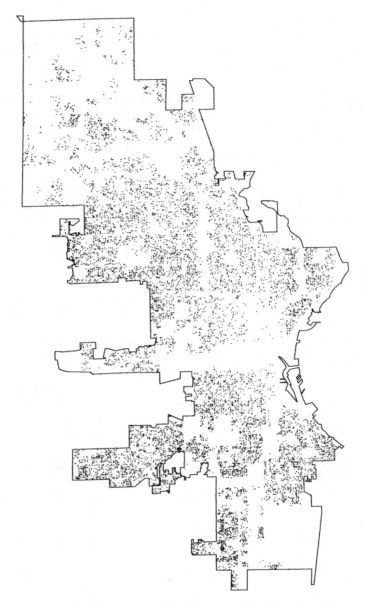

FIGURE 3.13. Citywide distribution of building permits.

a true figure for workload, then, the system must add the numbers
together rather than count the number of dots.

Since building inspectors depend heavily on the automobile in their
inspection activities, it was desirable to align the new district boundaries along major streets. Thus, as a guide in drawing the new districts,

census tract boundaries were overlaid onto the working map, as shown in Figure 3.14. (Census tracts usually have major streets or natural features for boundaries.)

Now the fun: With the GIS "digitizer," the supervisor drew a polygon around a portion of the city as a temporary district boundary and

FIGURE 3.14. Building permits with census tract boundaries overlaid.

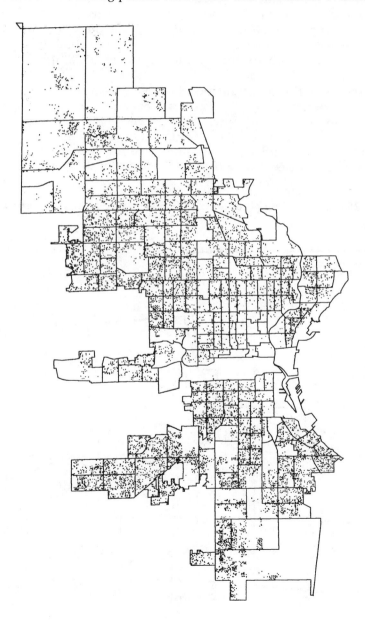

then invoked a program to add all the numbers within that area. The screen displayed the polygon on the map and the number of permits in that area. Figure 3.15 shows a few of these polygons with their totals for the northern part of the city.

Notice that the northernmost area contains 1201 permits, while the areas immediately south of it contain 1551 and 1329. To get them more in balance, it is a simple procedure to move the boundary they share down and then recompute the totals. This is how workload balancing works: change the polygon boundaries until the totals are approximately equal. It is an "interactive" process with the building-inspection supervisor making map changes and the GIS responding with new calculations.

In the case of these boundaries, however, the supervisor knows that the northern part of the city has large lots and a considerable amount of open space. Inspector travel time between buildings is greater here than in the more dense parts of the city. Therefore, the number of permits in these outlying areas should be fewer than the number in the inner-city areas. The 1201 should remain for the northern district.

FIGURE 3.15. Building permit totals for preliminary inspection district boundaries.

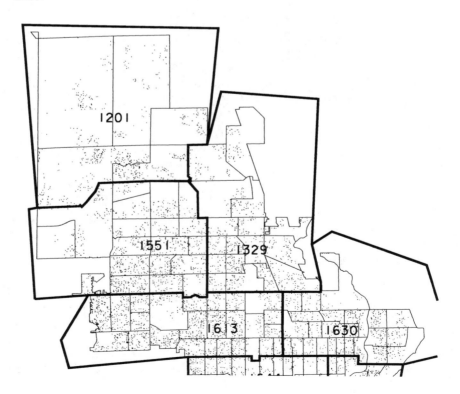

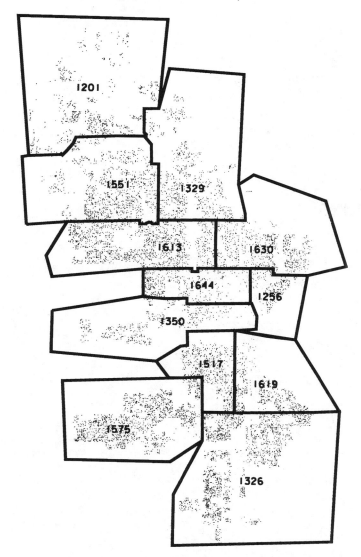

FIGURE 3.16. Building permit totals for final inspection district boundaries.

And so this interactive process continues: drawing a district boundary and looking at the total workload within the area. The supervisor uses his judgement in adjusting the boundary of the polygon and has the GIS recalculate new totals. The process takes seconds, and there are no census tract boundary restrictions as in the manual method—the boundaries can be placed wherever the supervisor wishes, and the totals are accurate for the new area. Figure 3.16 shows the polygons with new workload totals in each.

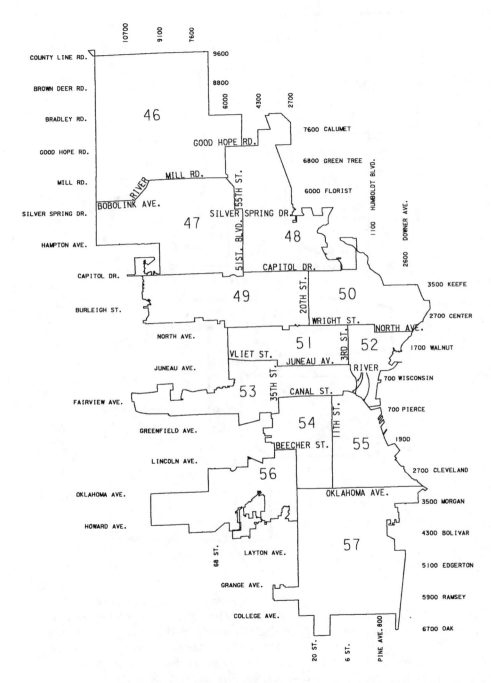

FIGURE 3.17. New building inspection districts with balanced workload.

Figure 3.17 is the final product: the new electrical inspection districts, balanced in terms of permit inspections (with travel time taken into account). It took 3 hours to set the boundaries and another hour to set up the final plot. The total time to balance the workload and redraw district boundaries was 4 hours, considerably less than the 480 work hours needed by 13 inspectors to build even stacks of permit cards.

Solid Waste Collection Routes

The city's Bureau of Sanitation is responsible for two primary services to the citizens: the collection of solid waste and the control and removal of ice and snow in the public way. Geographically, the Bureau has managed these services since 1970 in three service areas (North, Central, and South), each area in turn being subdivided into districts, resulting in a total of nine sanitation service districts for the city.

For solid waste collection, each of the nine service districts is further subdivided into individual collection routes (garbage routes, if you will). These routes were established in 1970, not as a network path for a garbage truck to follow, but as a geographic area within which one sanitation crew (a truck driver and two collectors) was expected to collect garbage in one week's time. Collection service is provided to all residential structures with four or fewer dwelling units (larger apartments and commercial structures were serviced by commercial firms). At the time, the crews provided "backyard" or "alley" pickup of garbage cans.

Solid waste collection routes, then, are geographic areas that define a standard work week for one sanitation crew. If a crew is efficient and completes its route in less time, then it is rewarded with time off (until the week is over and it is time to start at the beginning again). Conversely, if the crew is slow in collecting garbage, it is required to work extra hours to complete the route within the week.

In the 14-year period between 1970, when the routes were initially established, and 1984, two significant trends and one major policy decision occurred that affected the way solid waste collection was managed in the city. First, there was a gradual displacement of population from the center of the city outward, causing a decrease in workload of the inner city routes and an increase in workload of the routes on the fringes of the city. Second, a successful attempt to revitalize the city's central business district reduced the amount of downtown residential solid waste collection. Finally, the introduction of garbage carts, city-provided trash receptacles on wheels to replace individual garbage cans, promised to make the collection process easier and thus reduce the size of the crew from three to two (a driver and one collector). All of these changes affected the workload of the routes established in 1970.

The most dramatic of these changes, however, occurred in 1984 when city policy-makers decided to embark on a 6-year phasein of the "cart system," as it was called. By reducing the size of the crews, the Bureau was expected to reduce its staffing level from the 1000 employees it had in 1970 to 250 by 1990 when the cart system phasein was complete. When that decision was made, it was clear to Sanitation Bureau management that a major redefinition of solid waste collection routes was required. In addition, since the new program was expected to take 6 years to complete, some routes would be "cart routes" and some would be "can routes" during the phasein period, with more cart routes being added each year. A redefinition of all routes would be required in each year of the phasein period. The administration staff within the Bureau was faced with the monumental task of planning the new route boundaries each year for the next 6 years, knowing that adjustments would be required throughout the period as experience was gained on the new cart system.

The city's geographic information system was called upon to assist in this effort. Since one garbage cart was expected to be sufficient for one family per week, Bureau management decided that the size of a collection route would be determined by the number of dwelling units (i.e., carts) that could be processed in 1 week. That number, after considerable research, was determined to be approximately 2200 dwelling units. The task, then, was to establish new solid waste collection route boundaries that encompassed an area containing approximately 2200 dwelling units.

The solid waste collection route redistricting effort on the city's GIS was approximately the same as the efforts to reapportion voting wards and aldermanic districts and to redistrict building inspection districts, with the exception that the polygon processing would be applied to dwelling units instead of population and building permits. In addition, the only dwelling units to be processed would be those in single-family dwellings, two-family dwellings, and apartment buildings with four or fewer units.

The city's Master Property File, an indexed flat file containing many attributes of every property in the city, was used as the source of data for the redistricting process because it contained an attribute that identified the number of dwelling units in the building. It also contained the property number (tax key number) of the parcel on which the building was located. The tax key number of each record in the Master Property File having a Units attribute fewer than five was matched to the tax key number on the cartographic data base, and the value in the Units field displayed at the centroid of the property. This then allowed the polygon processing programs to perform calculations only on those properties that had four or fewer dwelling units and also calculate the total number of dwelling units in a polygon.

Polygons (new solid waste collection routes) were drawn on the system by Sanitation Bureau personnel who then invoked the polygon processing programs to calculate the total number of dwelling units in the area. If the total of the dwelling units was not close to 2200, then the operator adjusted the boundaries of the route and recalculated a new total. The process was repeated until all routes contained as close to 2200 dwelling units as possible. The result of this process is shown in Figure 3.18, which shows the projected route boundaries and route numbers when the cart system is completely phased in for the entire city.

Once the new collection routes were established, the crews needed maps of their new routes. These were produced by selecting the new route boundaries and overlaying the base map information on public right-of-way and street names, as shown in Figure 3.19. Maps such as these were taken out into the field to help guide the crews through their new assignments.

Public Policy—Housing

In most mature urban areas where growth is no longer a public policy issue, neighborhood stability is a leading concern of property owners whose single major investment in life is a home—the value of which, to a large degree, is dependent upon the quality of the surrounding neighborhood. A safe, clean, and aesthetically pleasing neighborhood is where people seek to live and where people want to live is where housing values are increasing or remaining stable.

Since vacant buildings and homes in disrepair are an eyesore to a neighborhood, they also are the "bad apples" that can erode the economic stability of a neighborhood. Public officials of large urban areas, aware of the gradual decay in neighborhood quality because of individual properties in disrepair or boarded up because of long-term vacancies, constantly seek innovative programs to prevent these conditions. Two such programs, the "Dirty Dozen" and the "Board-Up Survey," have been used in the City of Milwaukee for a number of years to combat neighborhood decay.

The "Dirty Dozen" Housing Code Violators

Annually, since 1985, one alderman has led a crusade against the top building code violators in the city: the "Dirty Dozen," as he calls it. The Dirty Dozen are the twelve property owners in the city who have had the most properties with building code violations during the previous year (see Figure 3.20). The alderman releases the list to the media every Spring not only to bring the issue to the public ("the City is watching you"), but also to publicly embarrass those who, he feels, "flout our laws in a variety of ways."

Whereas local building code enforcement efforts routinely slap fines on home owners who do not maintain their properties in accordance with established building code ordinances, many absentee landlords

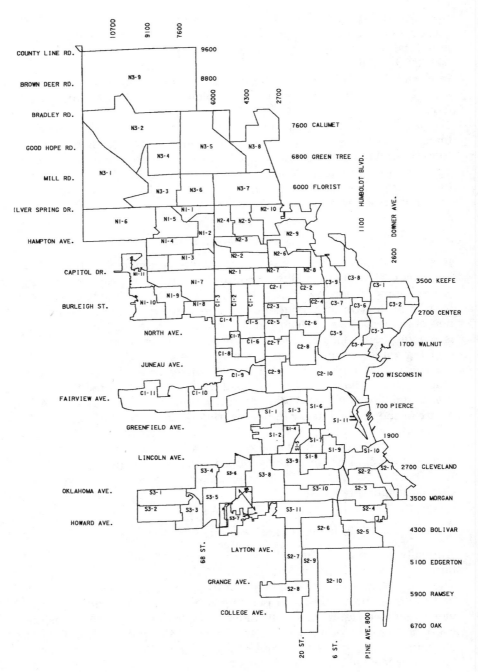

FIGURE 3.18. Projected solid waste collection "cart" routes.

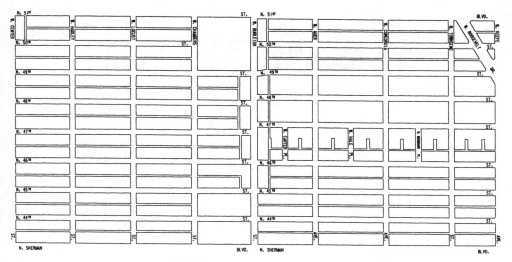

CENTRAL AREA I ROUTE 2

FIGURE 3.19. New solid waste collection route map.

amass a number of low-cost income-producing rental properties and find ways to circumvent the laws or delay the rehabilitation of their properties while the monthly rent checks keep coming in. The alderman stated at a recent press conference that "residents want decent, affordable housing, yet many families are forced because of need to accept living conditions that are indecent, unsanitary, and sometimes hazardous." The Dirty Dozen list shown in Figure 3.20 identifies the individuals who have the greatest impact on these people and the neighborhoods in which they live.

Figure 3.21 displays the locations of the properties owned by the Dirty Dozen that had building code violations during 1987 (they may own other properties that did not have violations, but they are not displayed on this map). Their spatial distribution identifies concentrated geographic areas impacted by only a dozen individuals. Not only does this map show the extent of their impact on the tenants, but the fact that they are concentrated in the near-northwest area of the city indicates that *all* the residents of these neighborhoods are affected by this handful of people.

The alderman noted, in his press conference, that "a very disturbing pattern is revealed by the number of repeat offenders. A property owner who has made the list two and three years in a row is obviously making no attempt to change (his or her) method of operation (to) make necessary repairs before orders are issued. This unconscionable method of operation is a drain on our city taxpayers, and our already overburdened Building Inspection Department must spend a great deal of time chasing these particular owners down."

City of Milwaukee

NEWS RELEASE

200 E. Wells Street, Milwaukee, WI 53202

For information call:

For Immediate Release: Thursday, July 28, 1988

Ald. Paul Henningsen or
Hettie White (278-2221)

LATEST "DIRTY DOZEN" CODE VIOLATORS REVEALED

NAME	ADDRESS	PARCELS OWNED	TOTAL 1987 VIOLATIONS	TOTAL UNABATED VIOLATIONS (AS OF JULY 15, 1988)
Jesse D. Hyche	11715 N. Silva	29	529	354
Lucy M. Hyche	Mequon	12	232	183
Mary T. Posnanski	5650 N. 64th St.	64	494	54
Joseph R. Peters	1870 N. Warren	71	479	253
Duncan LaPlant	3752 N. Teutonia	56	409	67
Marian LaPlant	3752 N. Teutonia	41	323	65
Gary M. McHugh	3850 N. Holton	57	372	243
Lawrence & Glenda Thompson	8990 N. Bethanne Dr.	19	317	263
John P. Savage	2120 W. Clybourn	76	301	67
Ernest Spaights	5275 N. Lake Dr.	36	289	122
Harold E. Dixon	5275 N. Lake Dr.	34	203	95
John D. Whitenack	2214-A N. 21st St.	28	289	169
Bernita Jacobson	2943 N. Maryland	8	256	149

FIGURE 3.20. The "Dirty Dozen" list of top building code violators.

One such repeat offender owns properties confined to one small neighborhood in the city, as shown in Figure 3.22: John Savage, whose 76 properties had 301 building code violations during 1987. "Mr. Savage," the alderman claimed, "owns over 50 properties scattered throughout a small neighborhood called Merrill Park. I believe the way Mr. Savage operates his properties . . . has been a major factor in preventing the revitalization of Merrill Park. This aspect of these top violator's methods of operations," he continued, "is particularly critical at a time when our city's leaders are focusing on neighborhood revitalization."

The city's geographic information system was used to match addresses between the Building Inspector's Housing Code Violation File and the Tax Assessor's Property Owners File to produce the list of the

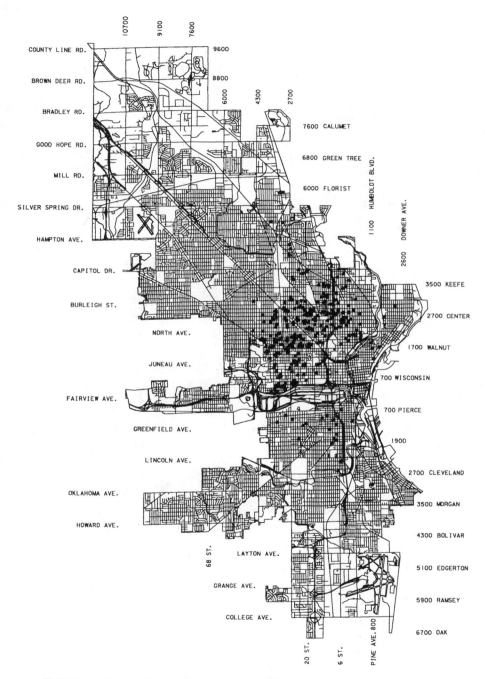

FIGURE 3.21. Locations of the properties owned by the "Dirty Dozen".

99

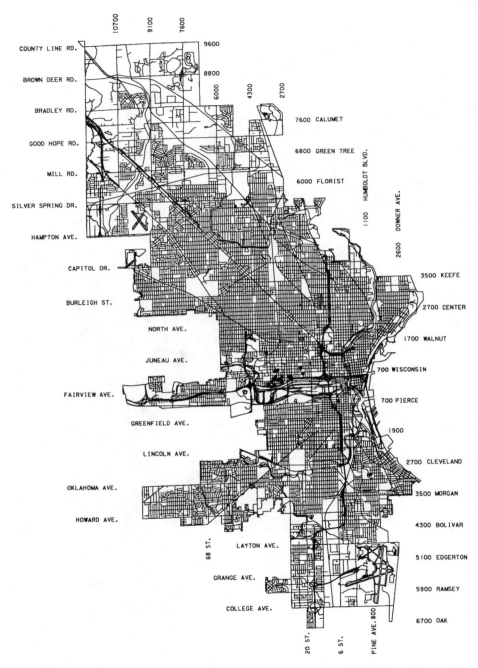

FIGURE 3.22. Locations of the properties owned by one of the "Dirty Dozen."

Dirty Dozen shown in Figure 3.20. The addresses of the properties with code violations were matched to their parcel numbers to access the map files and display the map in Figure 3.21. Figure 3.22 was produced by displaying the locations of the properties with code violations whose computer records contained an Owner Name of "John Savage."

Boarded-Up Properties

Some absentee landlords, feeling the public embarrassment of being a member of the Dirty Dozen, or merely computing the economics associated with the cost of correcting hundreds of building code violations, may feel that it is no longer in their best interest to continue landlord operations and, instead, board up and discontinue the rental of the property. Other owners, sometimes owner–occupants, may not be able to meet the mortgage payments and experience a foreclosure, in which a financial institution assumes ownership. If the property cannot be sold, the financial institution boards it up to prevent vandalism. Whatever the reason, boarded-up properties are an eyesore to a neighborhood and lead to neighborhood decay.

The city's Planning Department has been monitoring the status of boarded-up properties in the city since 1985 in order to identify trends and provide quantitative information to the elected officials for use in program and policy deliberations. This is accomplished through an annual residential board-up survey performed by Planning Department personnel in the Spring of every year. The survey is conducted within a special survey area comprising about 25 percent of the city. Every residential property in the survey area is viewed in order to determine whether or not it is boarded up. The city's geographic information system is central to this process.

The city's geographic base map is accessed to produce a series of maps for survey personnel. Each map covers a portion of the survey area that survey personnel can complete within two or three hours. The use of these maps allows the survey personnel to drive around the area and record the addresses of parcels with boarded-up buildings on them. Using the city's geographic data base in this manner assures that every parcel in the area has been surveyed and that no blocks have been omitted from the survey.

Once the survey is complete and all maps have been returned to the project manager, the addresses of all observed boarded-up properties are entered into a microcomputer data base. The data base consists of data about boarded-up properties identified in previous surveys. Matching the new "board-ups" to this data base identifies chronically boarded-up properties. A diskette containing a file of all the board-ups identified in the survey (along with a chronic board-up indicator) is then produced and transferred to the city's mainframe to be matched with the Master Property File to obtain current address, owner, geocode, and other property characteristics. These data are then printed out on listings showing board-ups by Census Tract (Table 3.1), Owner Name (Table 3.2), Address, and Aldermanic District. Notice in the Owner Name list of Table 3.2 that one of the Dirty Dozen, Gary M. Mc Hugh, owns nine boarded-up properties.

In order to identify and analyze trends associated with board-ups over the years, summary data are presented in chart form, as shown in Figures 3.23 through 3.25. These charts show that, while the total number of board-ups is decreasing (from 780 in January of 1986 to 680 in June of 1988), the chronic board-ups—those remaining boarded up for a year or more—continue to increase gradually (Figure 3.25). Additional statistics by Aldermanic District are produced for the elected officials to provide them with information about the district they represent.

The Board-Up File is finally passed to the city's geographic data base, where the parcel numbers are matched to the parcel numbers on the maps and a symbol is placed at the center of the parcel for each board-up. The map of the survey area shown in Figure 3.26 displays the spatial distribution of all boarded-up properties identified in the survey.

Table 3.1. City of Milwaukee listing of boarded-up buildings by tract.

Tract	Premise address	Taxkey	Land use	Standard owner name
23	4624 N 19th St	232-0620-000	8810	City of Milw Housing Auth
Total boardups = 1				
24	4707 A N 30th St	230-0290-000	8820	Bates Marion
	4707 A N 30th St	230-0290-000	8820	Bates Leon
Total boardups = 2				
26	4771 N 40th St	229-0707-000	8820	Helfer Sebastian J
	4570 N 41st St	229-9880-000	8810	Community HSG Presv Corp
Total boardups = 2				
39	3531 A N 38th St	268-1102-000	8820	No owner-name record found
	3623 N 38th St	268-1031-000	8810	
Total boardups = 2				
40	4143 N 44th St	248-0718-000	8810	
Total boardups = 1				
42	4560 N 29th St	230-0657-000	8810	Landry Wesley
Total boardups = 1				
43	4037 N Port Washington Av	243-0706-100	8810	Rudolph June K
	4037 N Port Washington Av	243-0706-100	8810	Rudolph Daniel
Total boardups = 2				
44	104 W Keefe Av	273-0921-100	8820	Sec of Housing & Urban Dev
	3908 N Port Washington Av	273-1542-000	8810	City of Milw

Geographic information systems are able to combine digital map information with attribute data associated with features that can be located on a map. The systems can then assist in the analysis of data in a spatial context to address issues and problems related to the service delivery, management, and policy-setting functions of government. The figures on the following pages exemplify some of the capabilities of GIS technology when color is used to assist in these analyses.

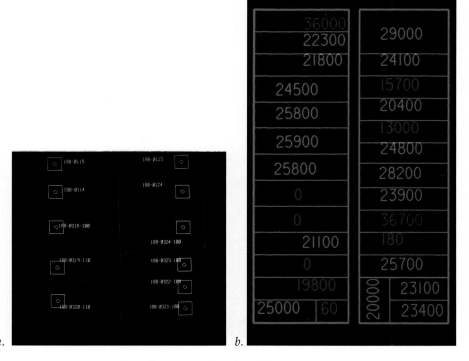

a.   *b.*

Figure 1. Location identifiers uniquely identify the location of specific features and also provide links between the features on a map and their attributes stored in data bases. (a) *Unique parcel numbers* (shown in blue) are stored on one layer of the digital map data base and are also links to a property data base (such as an Assessor's Tax Roll File). (b) *Property tax assessment values* can then be extracted from the data base, matched to their associated parcel numbers in the digital map data base, and displayed at the location of their parcels on the map. Color allows the classification of individual assessment values into ranges of values, each color representing a different range.

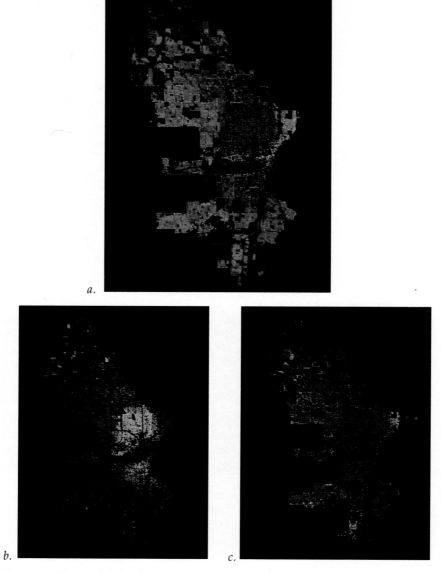

a.

b.　　　　　　　　　　　　　*c.*

Figure 2. *The spatial distribution of property tax assessment values* across an entire jurisdiction is enhanced by color coding the property centroids according to ranges of values. (a) Centroids for 160,000 parcels are color coded by value range with red representing values less than $20,000 and yellow representing values greater than $1,000,000. Other colors represent different assessment value ranges. Selected attribute values or ranges of values can emphasize the spatial distribution of properties sharing common characteristics by highlighting residential properties with (b) assessment values below $15,000 or (c) assessment values greater than $200,000.

Figure 3. *The age of buildings* can be analyzed spatially when the Year of Construction attribute in a data base is displayed for an entire jurisdiction. Patterns emerge when the Year of Construction attribute is displayed for all of the 160,000 properties in a city and the values are classified by color coded dots according to the decade in which the buildings were built. (This same technique has been used with crime data over one-week time periods to observe crime patterns over time and actually predict where future crimes are most likely to occur.)

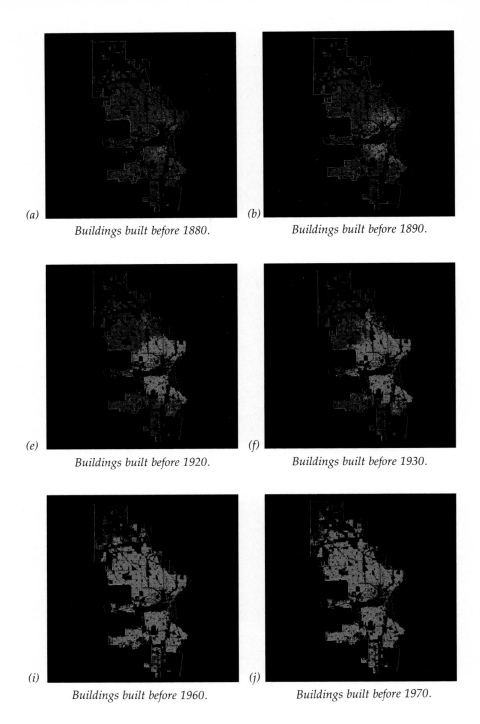

(a) *Buildings built before 1880.*

(b) *Buildings built before 1890.*

(e) *Buildings built before 1920.*

(f) *Buildings built before 1930.*

(i) *Buildings built before 1960.*

(j) *Buildings built before 1970.*

Figure 4. *Time* as an attribute of geographic data can identify spatial clustering resulting from events or happenings in the past to provide insights into current problems or issues related to urban governments. Highlighting the parcels

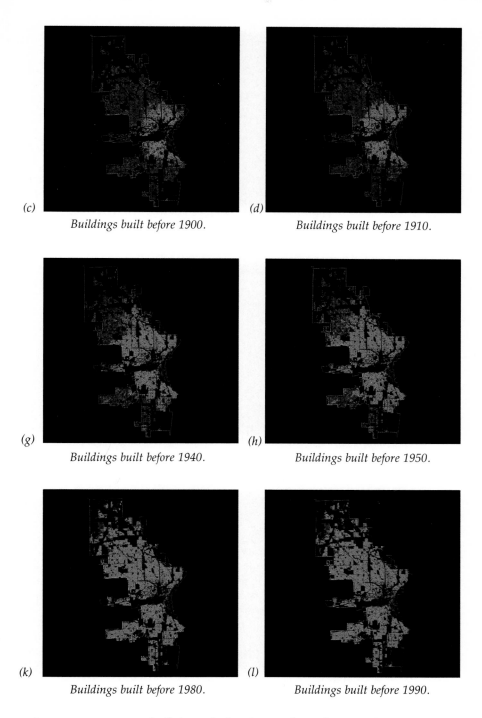

(c)   *Buildings built before 1900.*

(d)   *Buildings built before 1910.*

(g)   *Buildings built before 1940.*

(h)   *Buildings built before 1950.*

(k)   *Buildings built before 1980.*

(l)   *Buildings built before 1990.*

whose structures were built in each decade can show the growth of a city over a period of 110 years, beginning with the year 1880.

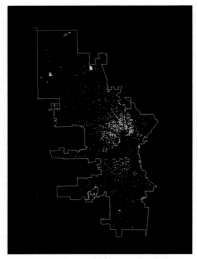

a.

b.

c.

Figure 5. The analysis of *property tax delinquencies* can be enhanced by combining data from different sources to investigate possible causes of unusual clustering. These maps identify unusual clusters of properties that are tax delinquent and show the overlay of additional map information to investigate possible causes for the clustering. (a) A dot density map of all property tax delinquencies shows two clusters in the northwestern portion of the city. (b) An enlargement of one of the clusters shows that the dots at this scale are not really dots, but actually are attribute values representing the number of years of delinquency. These properties have all been delinquent for 4 years. (c) An overlay of the land use layer of the digital map base shows that the lack of a land use symbol on the tax delinquent properties indicates that they are vacant lots. This could mean that they are tax delinquent because the developer is not paying taxes on them until they are built and sold. (Another overlay of the Owner Name attribute may confirm this.) The results of this analysis show that the properties are delinquent because the delinquency penalty is less than the prevailing interest rates, thus allowing the developer to use his money for other, more profitable investments. (City policy allowed up to 4 years of delinquency before legal action was taken.)

a.

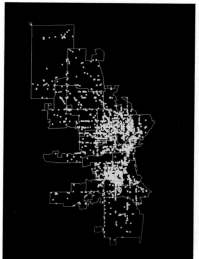

b.

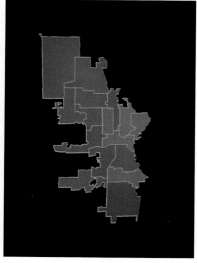

c.

Figure 6. The analysis of *liquor license applications* can be enhanced by using the polygon processing capabilities of a GIS to aggregate data from individual properties to a higher level of geography. These maps show where individual liquor licenses are located in the city and display the boundaries of aldermanic districts used for computing and displaying summary statistics. (a) Attributes of liquor licenses (Type of License and License Number) for two establishments are displayed on the property map layer of the digital map data base. This was accomplished by using the address of the establishment to find its parcel number and relating it to the parcel number on the digital map data base. (b) A dot density map showing liquor license locations for the entire city is produced by enlarging the area of display and turning off the layers containing property lines and addresses. The layer containing aldermanic district boundaries is then overlaid onto the map. (c) A choroplethic map of liquor license totals by aldermanic district is then produced by using polygon processing programs that count the number of licenses within each district and then shade the area of each district with a color that represents a range of totals. This allows the analyst to compare aldermanic districts on their liquor license distribution.

LESS THAN 75
76 TO 125
126 TO 200
201 TO 275
MORE THAN 275

Figure 7. *Overlaying liquor license locations onto a choroplethic map of aldermanic districts* allows the comparison of districts as well as the analysis of spatial distributions within districts. Since most socioeconomic occurrences are not related to political or administrative district boundaries, the use of choroplethic maps for some analyses can lead to misleading results. Often it is advantageous to investigate spatial clustering within the polygons that are being compared when using a choroplethic map. The choroplethic map of Figure 6c, for example, indicates a significant difference in total number of liquor licenses between aldermanic districts. This may cause two aldermen in adjacent districts to disagree on the impact of a new liquor license because one district may have many more than another. When the dot density map of Figure 6b is overlaid onto the choroplethic map, however, the distribution of licenses within the districts shows that some clusters really can affect more than one district. Aldermen representing adjacent districts may then agree on the impact of a new license in these areas.

Table 3.2. City of Milwaukee listing of boarded-up buildings by owner. Blank names indicate the same owner as the previous name.

Standard owner name	Taxkey	Premise address	Land use	Tract
Landry Wesley	282-1669-000	3283 N 5th St	8830	68
Lasher Barbara L	365-0947-000	1656 N 32nd St	8810	97
Lease Jean E	350-2070-200	2202 C N 25th St	8830	99
Lee Christine	323-0502-000	2519 N 6th St	8810	83
Lee Gusta	270-1903-000	3903 N 23rd St	8820	47
Leifer Bruce A	312-2302-000	3041 N 11th La	8810	66
Lemanczyk Frances	314-0791-000	314 E Hadley St	8820	81
Liptak Larry J	349-1045-100	2053 N 30th St	1796	98
Liptak Michael C	349-1045-100	2053 N 30th St	1796	98
Litza Bernard D	309-0803-000	2769 N 33rd St	8820	89
Lopez Tati	350-1336-000	1856 N 26th St	8810	120
Love Walter V	309-0626-000	3016 N 30th St	8810	63
Luy Jerome J	352-2043-000	2246 N 14th St	8810	102
Mack Greta	313-2328-000	2769 N 2nd St	8810	82
Mandli S G	433-0804-000	1329 S 15th Pl	8899	164
Marshall Frances E	313-1915-000	2822 A N 4th St	8820	83
Martin Dorothy	321-1588-000	311 E Clarke St	9999	81
Martin Franklin	321-1588-000	311 E Clarke St	9999	81
Matic Dusan	326-1461-000	2558 N 34th St	8820	89
Matthews Alma	351-1117-000	1905 N 19th St	8820	118
May Maxine R	284-0901-000	3285 N 16th St	8810	68
Mc Cain Roy	390-0137-000	1407 W Juneau Av	7022	146
Mc Cants Tommy	313-1607-000	231 E Burleigh St	8820	70
Mc Hugh Gary M	311-0409-000	1502 W Columbia St	8810	66
	350-2069-000	2472 B W Garfield Av	8820	99
	325-0091-000	2509 W Medford Av	8810	99
	283-0651-000	3122 N 11th St	8810	66
	283-0472-000	3261 N 12th St	8810	68
	352-1911-000	2111 A N 15th St	8820	102
	311-1887-000	2710 N 18th St	8810	86
	350-0430-000	1721 N 20th St	8820	119
	312-1794-000	2847 A N 8th St	8820	84
Mc Mcmillan Elma H	323-0502-000	2519 N 6th St	8810	83
McKenzie Gary O	349-0968-000	2909 W Vine St	8810	97
McNeal Frank	325-0057-000	2473 W Fond du Lac Av	1761	88
Ment Joel	365-1086-000	1521 N 33rd St	8830	122
Milde Rev Martha	270-1171-000	3627 N 25th St	8810	47
Miller Steven L	321-2245-000	1021 E Wright St	8830	107
Mitchell Maurice	310-1903-000	2526 W Center St	8820	88
Moore Clarence	322-0610-000	2637 N Palmer St	8820	82
Moore Johnel L	311-2801-000	2715 N Teutonia Av	8899	85
Moore Richard C	311-2801-000	2715 N Teutonia Av	8899	85
Moore Robert	287-1509-000	3403 N 39th St	8820	48
Moore Robert L	322-0610-000	2637 N Palmer St	8820	82
	310-0842-000	3078 N 22nd St	8820	64
Muhammad Lateefah	325-0035-000	2407 W Fond du Lac Av	7021	99
Muhammad Naim	325-0035-000	2407 W Fond du Lac Av	7021	99
Nash Elaine	312-3312-000	3043 N 12th St	8820	66

BOARD-UP SURVEY TOTALS

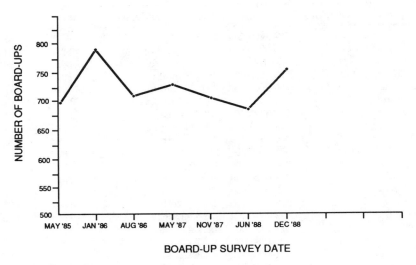

FIGURE 3.23. Graph of board-up survey totals by date of survey.

FIGURE 3.24. Graph of new board-ups by date of survey.

NEW BOARD-UPS

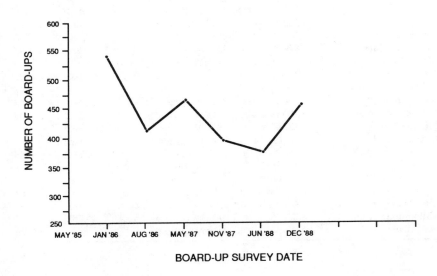

CONTINUED BOARD-UPS

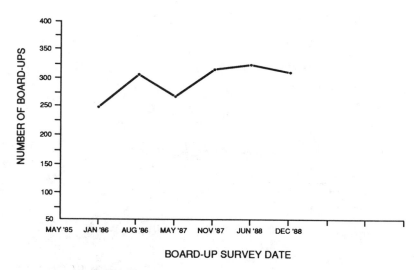

FIGURE 3.25. Graph of continued board-ups by date of survey.

Tax Delinquent Properties

In most urban communities, property tax delinquencies can also contribute to neighborhood decay because owners who do not pay their property taxes are less likely to invest in the cost of maintaining those properties, allowing the buildings to deteriorate. Over time, the unpaid taxes build up, and penalties are added. The cost to correct deficiencies identified in housing code inspections adds to the financial burden of the owner. These costs ultimately compound to the point at which the financial burden is greater than the value of the property itself and the owner "walks away" from ownership (allowing ownership to revert back to the city) to eliminate the accrued financial burden. Public officials do not want this to happen because, first, no revenue is produced from city-owned properties; second, uncollected property taxes cause higher taxes to be paid by the responsible property owners (to make up for the lack of revenue from tax delinquencies); and third, there is a cost to the city for rehabilitating these properties so that they can be sold and returned to the tax rolls.

Elected officials seek many avenues to prevent the process described above from occurring: They seek changes in state laws that allow the city to act more quickly and strictly with tax delinquent property owners; they apply pressure on the court system to ensure that recalcitrant tax payers are dealt with swiftly and sternly; and they work with other city officials to initiate administrative changes and new programs to

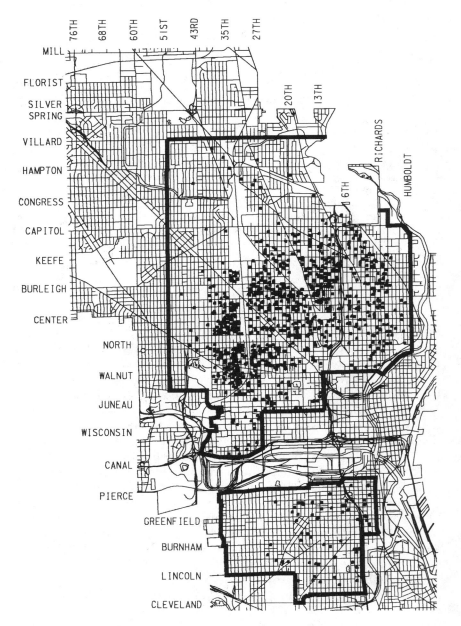

FIGURE 3.26. Locations of buildings that are boarded up.

106

address the tax delinquency problem. All of these efforts, however, may not be worthwhile if the extent of the problem is not known. The identification of the problem and its characteristics is a function a geographic information system can perform to aid these public officials in their efforts.

Figure 3.27 shows the locations of all 10,016 tax delinquent properties in the city as of March, 1986, with the boundaries of the aldermanic districts overlaid. It was produced by extracting all the addresses from the City Treasurer's tax delinquency data base and matching them to parcel numbers so that symbols could be placed at their appropriate map locations in the geographic data base. This map shows the spatial distribution of all tax delinquent properties throughout the city as well as throughout each aldermanic district, identifying neighborhoods in which the problem is most severe.

Table 3.3 gives summary data about these tax delinquencies by aldermanic district. While these statistics provide the quantitative information needed to evaluate the extent of the problem, they do not offer enough flexibility for further analyses to investigate whatever unknown geographic factors may assist in developing solutions to the problem.

Table 3.3. Property tax delinquency totals by aldermanic district as of March, 1986

District	Total tax delinquent properties	Total delinquent amount	Average amount delinquent per property	Percent of total delinquent amount
1	1256	$1,752,331	$1,395	9.77
2	466	677,922	1,455	3.78
3	385	906,306	2,354	5.05
4	781	1,841,020	2,357	10.26
5	171	294,776	1,724	1.64
6	1420	2,325,468	1,638	12.96
7	550	783,084	1,424	4.36
8	394	602,100	1,528	3.36
9	380	879,615	2,315	4.90
10	1245	2,000,379	1,607	11.15
11	168	278,841	1,660	1.55
12	741	1,744,454	2,354	9.72
13	283	682,568	2,412	3.80
14	328	613,377	1,870	3.42
15	697	1,444,498	2,072	8.05
16	534	908,532	1,701	5.06
Not Coded	217	206,484	952	1.15
Total:	10016	$17,941,764	$1,791	100

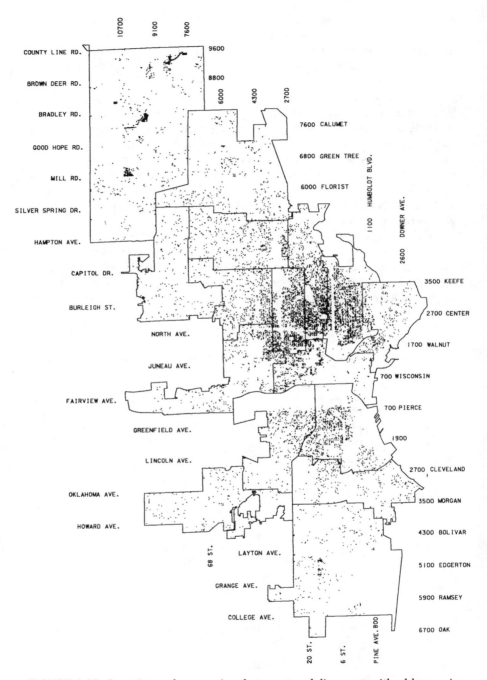

FIGURE 3.27. Locations of properties that are tax delinquent with aldermanic district boundaries overlaid.

Figures 3.28 through 3.30 allow this geographic analysis. Figure 3.28 is a choroplethic map showing the number of tax delinquent properties in each aldermanic district. The shading in each district represents one of five different classifications, as shown in the legend in the lower left-hand corner of the map. The more dense the shading of a district, the

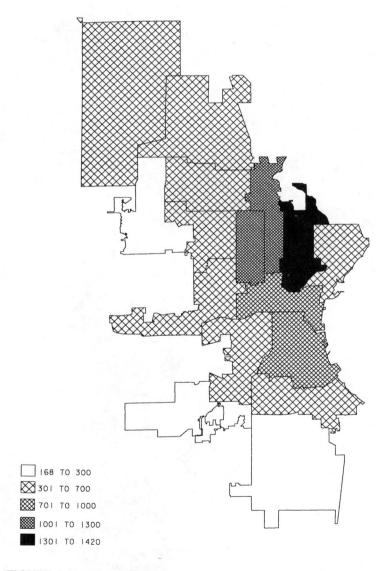

168 TO 300
301 TO 700
701 TO 1000
1001 TO 1300
1301 TO 1420

FIGURE 3.28. Choroplethic map of properties tax delinquent by aldermanic district.

more tax delinquent properties there are in that district. Notice that they are concentrated in 3 of the 16 districts (Districts 1, 6, and 10). Figure 3.29 compares districts by the average dollar amount delinquent per property. Different districts have the highest average amount of delinquency per parcel (Districts 3, 4, 12, and 13). Thus, one set of districts has the most delinquent properties, and a different set of districts has the highest average amount of delinquency per parcel. Two possible strategies can be developed from these maps: Targeting delin-

FIGURE 3.29. Choroplethic map of the average tax delinquency amount per property by aldermanic district.

$1395 TO $1500

$1501 TO $2000

$2001 TO $2300

$2301 TO $2500

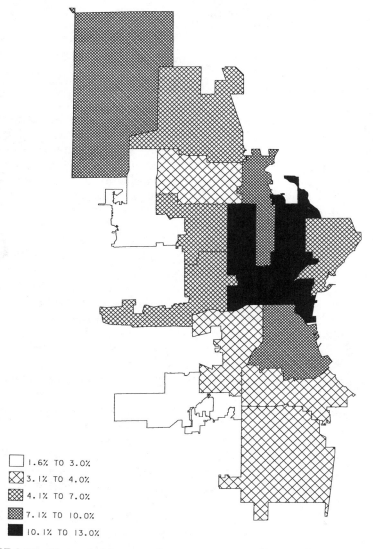

1.6% TO 3.0%

3.1% TO 4.0%

4.1% TO 7.0%

7.1% TO 10.0%

10.1% TO 13.0%

FIGURE 3.30. Choroplethic map of the percent of the total amount delinquent in property taxes by aldermanic district.

quent tax payment programs in the first set of districts (1, 6, and 10) may have the greatest impact on the most citizens because more properties are involved, while targeting such programs in the second set of districts (3, 4, 12, and 13) may be more productive because there would be fewer court cases to conduct. Figure 3.30 portrays a different view of the data: the percent tax delinquent amount that each district is of the total amount citywide. Notice that the top districts (Districts 4, 6, and 10) have almost 35 percent of the total city tax delinquent amount.

Each district is also in one of the highest groups of the previous maps in Figures 3.28 and 3.29.

Liquor License Policy

One of the most politically sensitive responsibilities of aldermen in the City of Milwaukee is that of the review and approval of liquor license applications and renewals. On one hand, responsible and successful tavern owners may not only provide an economic asset to the neighborhood, but they also can generate significant local support at re-election time for the alderman who represents the district. On the other hand, constituents who associate local problems (crime, noise, overcrowded parking, etc.) with these establishments also apply pressure on their alderman to eliminate what they perceive as the cause of these problems: the tavern.

In Milwaukee, the liquor license review process is conducted by the Utilities and Licenses Committee, a subcommittee of the Common Council. With over 2000 retail liquor licenses to review each year, the Utilities and Licenses Committee generally defers to the wishes of the alderman of the district in which the tavern is located when a license is reviewed. If the alderman has received no complaints from area residents, a liquor license can be approved with little opposition. Complaints by neighborhood residents about problems associated with a particular tavern (including complaints from the Police Department) cause the alderman to look more closely at the conditions and balance the positive and negative effects of the approval before making a recommendation. The city's geographic information system is valuable during this process.

Liquor license information (address, type of license, owner name and address, etc.) is recorded and processed by a computerized data base in the City Clerk's office. When a license is due for renewal, a notice is sent to the owner, who submits an application and the processing fee. When the renewal of a tavern's license is in question, and before it is on the agenda of the Utilities and Licenses Committee, the alderman from the tavern's district will often request data from the city's GIS to use in the analysis. Figure 3.31, for example, is a map showing the individual locations of all retail beer licenses in the city for 1985. It was created by matching the addresses in the liquor license data base with the city's Master Address Index to identify their parcel numbers, which then can be located on the digital base map of the GIS.

While this map provides a general view of the spatial distribution of retail beer establishments in the city, it does not provide any quantitative characteristics about this spatial distribution. Liquor license *density* (in licenses per square mile) is a statistic normally requested by aldermen for further analysis of spatial distribution. The automated mapping capabilities and the polygon processing capabilities of the GIS are

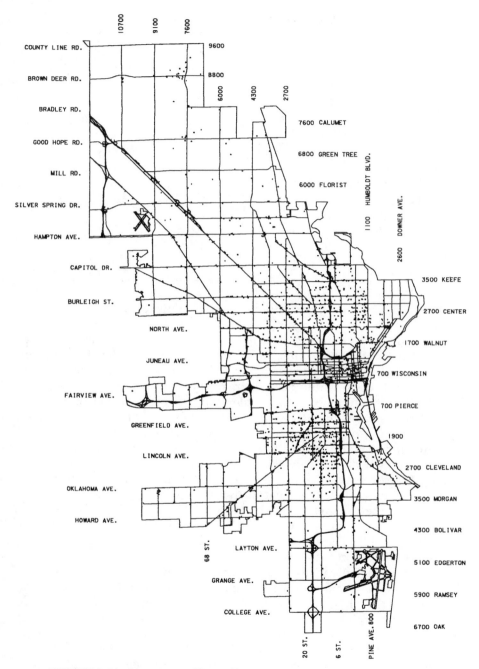

FIGURE 3.31. Locations of liquor licenses with major streets overlaid.

113

Table 3.4. Retail beer license totals by aldermanic district.

District	Area (sq. mi.)	Licenses	Licenses (per sq. mi.)
1	3.5	123	35.1
2	4.9	66	13.5
3	3.7	149	40.3
4	3.5	244	69.7
5	5.7	38	6.7
6	4.0	200	50.0
7	3.8	67	17.6
8	5.3	166	31.3
9	9.6	57	5.9
10	2.7	146	54.1
11	5.4	42	7.8
12	5.5	363	66.0
13	11.3	80	7.1
14	4.9	134	27.3
15	17.6	33	1.9
16	5.6	111	19.8

used to produce the liquor license density statistics as shown in Table 3.4.

The automated mapping capabilities of the GIS provide a simple means to calculate the square mile area of each district, while the polygon processing capabilities provide the counts of licenses (in the case of Figure 3.31, dots) within each district. The statistics in Table 3.4 show a significant difference among aldermanic districts in their density of retail beer establishments, ranging from a low of 1.9 taverns per square mile to a high of 69.7 taverns per square mile. From these statistics, it would appear that Districts 5, 9, 11, 13, and 15 could absorb additional taverns much more easily than Districts 4, 6, 10, and 12. (However, since District 4 includes the city's central business district, one would expect a higher density in that district.) Thus, an application for a new license in the low-density districts (5, 9, 11, 13, and 15) is easier to approve than an application for a license in the high-density districts (6, 10, and 12). The GIS displays this reasoning in choroplethic map form, as shown in Figure 3.32.

Since aldermanic districts are not necessarily homogeneous in their socioeconomic composition and vary in the geographic distribution of taverns within their boundaries, some aldermen prefer a more detailed geographic analysis of liquor license density. They feel that approving a license merely because it lies within the boundaries of a low-density district is not a rational decision if the new tavern is to be located close to most of the existing taverns in the district. Smaller areas within districts that have a low liquor license density may not be able to absorb an additional tavern without adverse reaction from local residents. When this is the case, an alderman will ask for a map such as that shown in

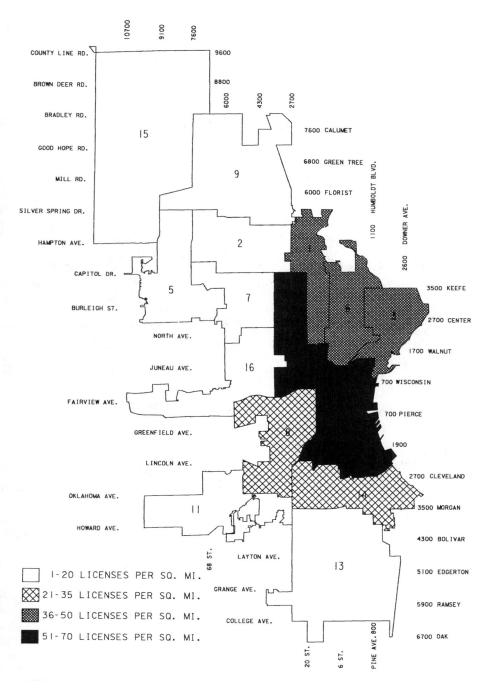

FIGURE 3.32. Choroplethic map of beer license density per square mile by aldermanic district.

115

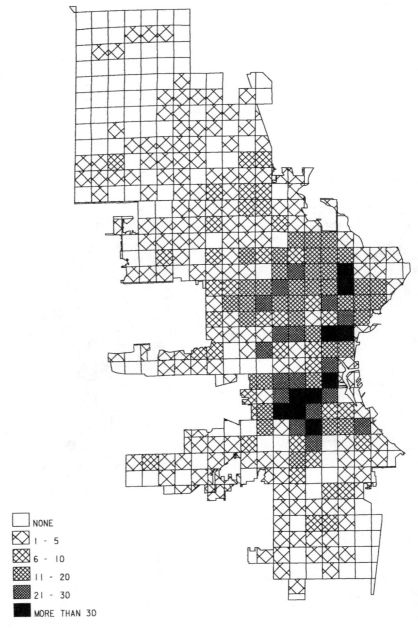

FIGURE 3.33. Choroplethic map of liquor license density per square mile by quarter section.

Legend:
- NONE
- 1 - 5
- 6 - 10
- 11 - 20
- 21 - 30
- MORE THAN 30

Figure 3.33, which displays liquor license density by quarter-section (½ mile square).

Health Policy on Lead Poisoning

Early in 1987, city Public Health officials observed an alarming number of lead poisoning cases reported in the city during the period of September through December of 1986. Health Department records for that period identified the home address of each person affected by lead poisoning as well as some socioeconomic data about the patient. This information was processed by the city's geographic information system to help in the analysis of the problem. It was hoped that a review of the spatial distribution of the cases would help in the analysis. The resulting map of lead poisoning cases is given in Figure 3.34.

In addition to the obvious spatial concentration of the reported lead poisoning cases in the north-central portion of the city (see the study area boundary in Figure 3.34), the socioeconomic data also identified another pattern: Almost all of the reported cases in the study area were children younger than 7 years of age.

Lead poisoning in children is characterized by mental retardation and brain damage. Federal researchers at the Center for Disease Control (CDC) in Atlanta have concluded that a tiny amount of lead—the equivalent of ten millionths of a gram of the toxic material in 3 ounces of blood—can cause these disabilities. Further research by the CDC found that chips of paint with a lead concentration greater than 50 percent were the largest source of lead in a child's environment. Toddlers who put paint chips in their mouths (they taste sweet at that age) comprised the majority of children who were poisoned each year. Other significant sources of lead are: soil and house dust, drinking water, food, automobile fumes, and industrial emissions. Lead poisoning, it would appear, is certainly an urban problem—one that could affect twelve million children under the age of 7 nationally because they live in homes with lead-based paint and thus are exposed to the largest source of lead in their environment. Nationally, 20 percent of urban inner-city children younger than the age of 7 suffered from lead poisoning. Doctors now believe that lead poisoning is a bigger health problem than asbestos or radon.

In Milwaukee, health-threatening amounts of lead were found in the blood of 10.8 percent of 5173 young children tested. Most of these children lived in older homes where they were exposed to lead in paint. Prior to 1950, paint with as much as 50 percent lead content was commonly used in residential construction. After that date, city building codes were changed to prevent the use of paint with such high concentrations of lead. The local problem with lead poisoning, it was concluded, was concentrated in homes built before 1950 that had children living in them who were younger than 7 years of age. The city Health

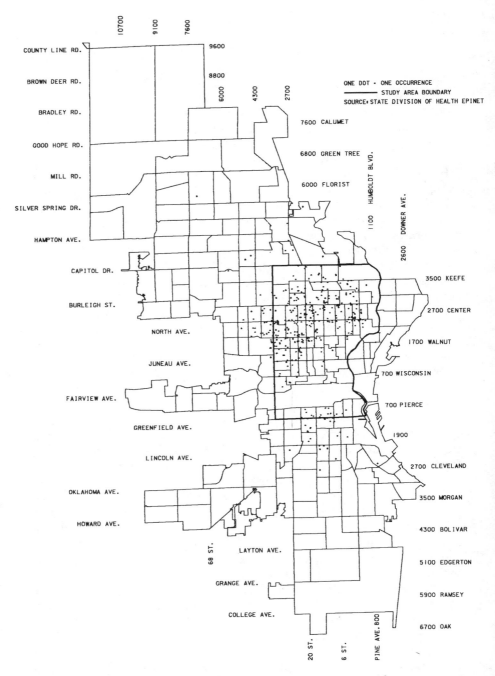

FIGURE 3.34. Locations of reported lead poisoning cases with census tract boundaries overlaid.

118

Department called for a major housing renovation program that would target older homes with peeling paint. Without such a program, department officials warned, several hundred children each year would be condemned to suffer lead poisoning unless lead-based paint was removed from hundreds of homes and apartments in the central city.

Hundreds of homes and apartments? How did the officials know the extent of a possible lead-based paint abatement program? Lacking detailed records of each child under the age of 7 living in a home built before 1950, the city's GIS was used to assist in the analysis so that an estimate of the extent and cost of such a program could be determined. Population data by age were available on a census tract basis from the annual school census conducted by the public school administration, but there were no data on the age of the home in which they lived. City tax records contained building characteristics for all the properties in the city with address, year built, type of structure, and so forth, but had no data on the socioeconomic characteristics of the residents of the buildings. To combine these two sets of data, a bivariate analysis was performed, using age of population by census tract as one variable (see Figure 3.35) and the number of homes built before 1950 (tax records aggregated by census tract) as the second variable (see Figure 3.36). Census tracts having the most children under the age of 7 and having the most homes built before 1950 would define the areas of the city in which a lead-based paint abatement program would have the greatest impact.

The result of combining the maps in Figures 3.35 and 3.36 is shown in Figure 3.37. Combining the two variables by census tract shows tracts with low concentrations of children under the age of 7 (0–234 children) and low concentrations of homes built before 1950 (0–203 homes). Those tracts have no shading in Figure 3.37 because they would be the least likely areas to realize an impact from an abatement program. Those tracts with the highest concentration of children under the age of 7 (389–1255 children) and the highest concentration of homes built before 1950 (525–1382 homes) would be most likely to realize the greatest impact from the abatement program; so they have the darkest shading in Figure 3.37.

Looking at this most darkly shaded group of census tracts made it clear that it was not "hundreds" of homes that required paint removal, but tens of thousands of homes. The cost of removing, shipping, and burying all the lead-based paint from each of the homes in these census tracts would require millions of dollars. (Nationally, it was estimated that the cost to eliminate the hazard in all public and private housing units with these characteristics would approximate $200 billion.)

The city scaled back its lead-based paint abatement program and chose, instead, to eliminate the hazard in all of its 2000 public housing units. It was one of the first cities in the country to establish an abatement

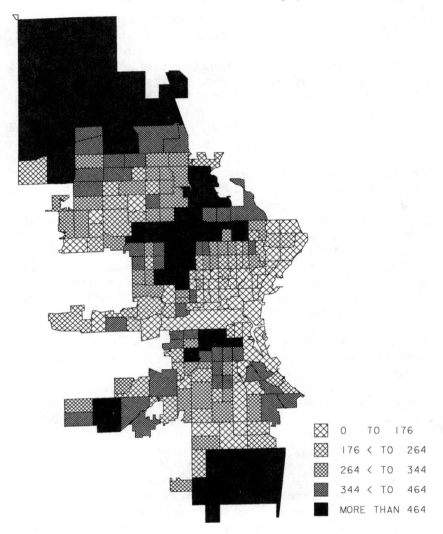

FIGURE 3.35. Choroplethic map of the number of children under the age of seven by census tract.

program, spending about $1.5 million of its $8.5 million allotment from the U.S. Department of Housing and Urban Development public housing modernization grant.

Library Facilities Planning

Library officials felt that there was a need for an additional neighborhood library on the northwest side of the city. With four neighborhood

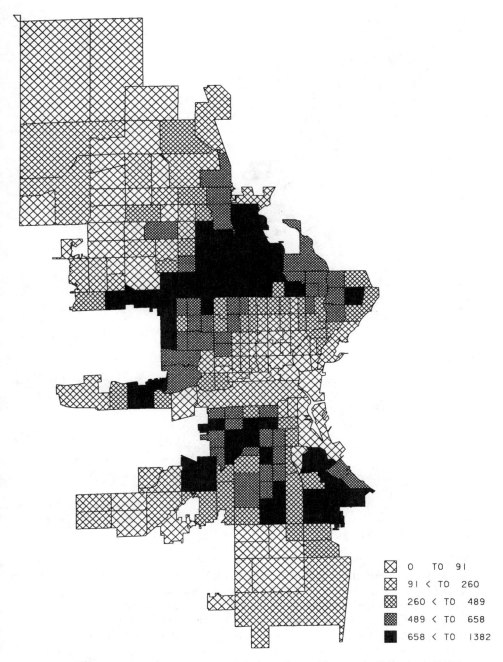

⊠	0 TO 91
⊠	91 < TO 260
⊠	260 < TO 489
▨	489 < TO 658
■	658 < TO 1382

FIGURE 3.36. Choroplethic map of the number of homes built before 1950 by census tract.

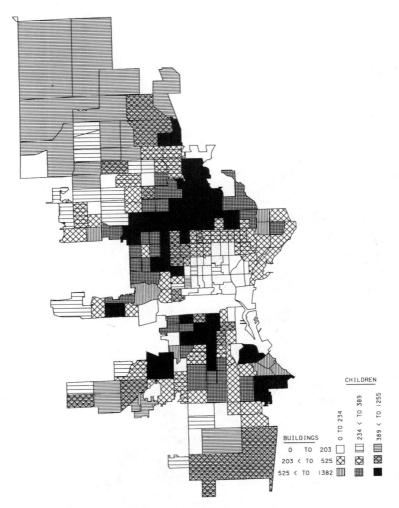

FIGURE 3.37. Bivariate choroplethic map of children under the age of seven and homes built before 1950.

libraries and one extended service library already in the area, they were not convinced of the need and asked the city's Planning Department to study the situation and develop a recommendation. They had previously determined that the standard service area for the types of services provided by a neighborhood library was based on the size of the population in the library's immediate geographic area. Typically, a neighborhood library can service a population of approximately 60,000 people, and an extended service library, because of additional services available, can serve approximately 70,000 people.

The city's geographic information system was used by the Planning Department to study the need for an additional neighborhood library.

The approach selected was first to locate the five existing libraries and then to determine their service areas based upon the population standards given by the library officials. Determining the size of the population lying outside those service areas would then not only determine the need for a library but would also give an indication of where such a library should be located. The city's population data base, containing population counts for each block in the city, would be used on the GIS to calculate the population served by each library.

Figure 3.38 displays the result of the analysis. The location of each of the five libraries is identified by a symbol consisting of three concentric circles. The GIS operator then created a circle on the work station screen using each library's location for its center and a point some arbitrary distance out as the end of its diameter. The system then drew

FIGURE 3.38. Service areas of neighborhood libraries based upon a surrounding population of approximately 60,000 people.

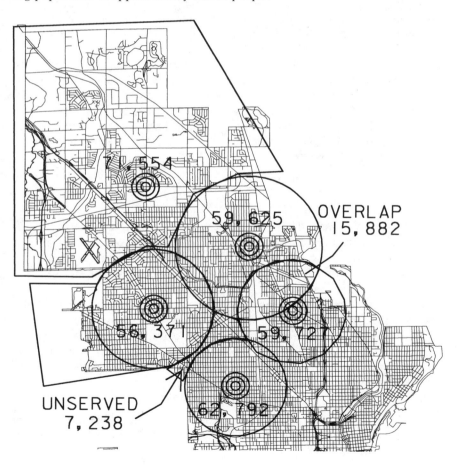

the circle and identified all of the blocks that lay inside. With the data base of population counts by block linked to the map by block number, the system summed the population counts of all the blocks inside the circle and displayed the total on the screen. If the total was significantly less than 60,000 (70,000 for the extended service library), the circle was deleted and recreated with a larger diameter. If the total was significantly greater than 60,000 (70,000 for the extended service library), the circle was made smaller. This interactive process continued until circles for each library defined areas having total populations as close to 60,000 (or 70,000) as possible.

Notice that Figure 3.38 shows two results from this analysis: Some circles overlap, and some areas of the city are not included in a circle. This meant, of course, that in some areas there was an overlap of service areas between two libraries, and in other areas some city residents were not being served by any library. The next step was to determine the size of the unserved population.

To determine the size of the population not served by a library, the GIS operator digitized a polygon of the area not included in any circle. The population counts for the blocks within that polygon were summed and displayed as a total "unserved" population of 7,238 residents. The "overlap" population served was determined similarly and totaled to 15,882 residents. Thus, the number of residents who were not served by a neighborhood library (using a service area standard of 60,000 people) was only 7,238—far below the standard. Furthermore, there were 15,882 residents who were served by more than one neighborhood library. Clearly, there was no need to spend the taxpayer's money on a new library on the northwest side. No library was built.

Exercise

Pick up a newspaper. There are reports almost daily of local issues or problems that require information related to geographic locations in a jurisdiction: the racial composition of school board members; minority representation on jury duty; automobile insurance rates higher in some areas of the city than others; response time for emergency vehicles; problems with obtaining home mortgages in the inner city; urban flight; site selection for a new ball park; property tax reassessments; arson and crime waves; the effects of floods and drought on property. Select one of interest that has recently been publicized in your city and prepare a paper describing how an urban GIS can assist in resolving the problem or issue. Address the following topics:

1. *Description of the problem or issue*
    a. Briefly state the problem or issue in everyday terminology that the average taxpayer could understand. What are its effects on the citizens?

b. Now state the problem or issue as though you were proposing the use of a GIS to assist the top manager responsible for resolving the problem or issue. How can GIS technology help operations, management, or planning responsibilities? Key words will be: how many, where, what kinds, how much, etc.

2. *Approach to solve the problem or address the issue*
   a. How would the organization normally handle this issue or problem without the benefit of a GIS? (Remember: most organizations have computerized data, but many do not have a GIS. Is it merely a matter of assigning a lot of people to do work, or is there more required?)
   b. If you were a manager of a GIS, how would you use the resources available from the system to solve the problem or analyze the issue?

3. *GIS resources required*
   a. Using the list of GIS tools given in Chapter 2, what is needed in the GIS to address the problem or issue?
   b. What points, lines, and/or polygons must be defined?
   c. Which data bases must be available?
   d. To what degree does accuracy affect the problem?

4. *Expected results*
   a. Simulate a GIS. Assume that the resources are in place and that you can use the system to assist the manager to address the issue. What steps would you take to address the issue with a GIS?
   b. Create examples of expected outputs: maps, tables, listings, screen displays, etc.

Additional Readings

Godschalk, David R., Bollens, Scott A., Hekman, John S., and Miles, Mike E. (1985), *Land Supply Monitoring—A Guide for Improving Public and Private Urban Development Decisions*, Oelgeschlager, Gunn & Hain, Publishers, Inc., Boston, Mass.

Huxhold, William E., (1987), "Modernizing Land Information Systems for City Planning and Management: Problems and Opportunities," *Geomatics Applied to Municipal Management*, Canadian Institute of Surveying and Mapping (Montreal Branch), Montreal, pp. 353–76.

Huxhold, William E., Gschwind, Randolf, and Allen, Richard (1982), "An Evaluation of the City of Milwaukee Automated Geographic Information and Cartographic System in Retrospect," *Harvard Computer Graphics Week*, Harvard University, Cambridge, Mass.

Martin, James (1976), *Principles of Data-Base Management Methodologies*, Prentice Hall, Englewood Cliffs, N.J.

Descriptions of additional urban applications of geographic information systems can be found in papers contained in the following publications:

ACSM, ASPRS, AAG, and URISA (1988), *GIS/LIS '88 Proceedings*, Washington, D.C.

Joint Nordic Project, *Digital Map Data Bases—Economics and User Experiences in North America* (1987), Nordisk Kvantif, Board of Survey, Finland.

National Computer Graphics Association (1987), *NCGA's Mapping and Geographic Information Systems '87 Proceedings*, NCGA, Washington, D.C.

Niemann, Bernard J., Jr. (ed.) (1984), *Seminar on the Multipurpose Cadastre: Modernizing Land Information Systems in North America*, Wisconsin Land Information Reports: Number 1, University of Wisconsin - Madison, Madison, Wisc.

Urban and Regional Information Systems Association, *URISA Annual Proceedings*, (each year 1980–89), URISA, Washington, D.C.

4

Topological Data Structures

To understand the intelligence topological data structures add to a cartographic data base in a geographic information system, it is useful first to look at how a geographic information system stores and processes its cartographic information. By understanding the digital representation of map information in a cartographic data base, one can observe how a map is generated and then how topological data structures enhance the use of the information for analysis and manipulation.

Graphical Representation of Cartographic Data

A geographic information system stores map information in digital form by using a cartographic data base consisting of individual records that contain attributes of each *cartographic feature* represented on a map. A cartographic feature is something that can be named (such as a street or manhole) and can be located on a map as well as on the ground in the real world. When it is seen on the ground, it is a *cartographic entity*. When the attributes that describe its graphic representation are stored digitally, it is a data record called a *cartographic object* (Moellering, 1988). Thus, the cartographic feature "street" is not only the entity upon which we drive our cars, but it is also that object that is stored in the computer and displayed on a map. The automated mapping functions of a geographic information system transform the attributes of cartographic objects stored in the cartographic data base into graphic representations on a map.

Cartographic data—consist of the attributes of cartographic objects used by the automated mapping software of a GIS to display map information.

Chapter 2 introduced the rule that anything of interest on a map must be *explicitly* defined in order for a geographic information system to process and display it. This is because a map is not stored in the cartographic data base of a geographic information system in the same form as that seen by the human eye on a sheet of paper or on a work station screen. In the computer, maps are stored in a cartographic data base that contains a record for each cartographic object, line of text, or symbol. These records contain attributes that describe to the system how to process and display the features to produce a map. Each type of feature—a point, line, polygon, text string, or symbol—has attributes unique to that feature. A line, for example, has a beginning point and an ending point, a thickness (line weight), color, etc., while a text string has a font style, a height for the letters and numbers, a point where it is displayed, an angle that the text follows, and other characteristics. Typical cartographic records for displaying these features are given in the data-base schema of Table 4.1.

Notice that the Area Record of Table 4.1 contains an attribute called "Vertices." This is because the system requires complete closure of the boundaries of polygons so that geometric calculations (such as computing the area inside the polygon) will not find any gaps between the lines that form its boundary. In an automated mapping system, a line feature (such as a street segment) that is also the boundary of a polygon (such as a census tract) is treated as two separate records because there is no topological definition relating the line to the polygon. The Symbol Record also contains information about lines. That is because a symbol is defined as a generic object that can be placed at any location on a map (a symbol for a bridge, for example, will look the same on a map wherever it is placed). The attributes of the lines that form the shape of the symbol must be defined in this record in order for the object to be plotted as a symbol. All of the records in Table 4.1 contain an attribute called "Level." Sometimes called a "coverage" or a "map

Table 4.1. A typical data-base schema for cartographic records with the attributes needed to plot cartographic features on a map.

Point record	Line record	Area record	Text record	Symbol record
Symbology	Symbology	Symbology	Font Type	Name
Color	Weight	Color	Height	Color
Level	Color	Level	Angle	Line Attributes
XY Coordinates	Level	Verticies	Color	Level
	Beginning Point		Level	XY Coordinates
	Ending Point		Beginning Point	
			Value	

facet" or some other term unique to specific commercial systems, the Level attribute identifies on which overlay the feature is. A map contains several overlays, each containing its own homogeneous set of data (such as property lines, street lines, street names, zoning boundaries, etc.), as will be described in Chapter 6.

Since there are many commercially available geographic information systems in use at present, each having its own unique design for processing features to be displayed on maps, the attributes in the records of Table 4.1 should be considered examples of the types of attributes needed for display and not universal specifications for all vendors. Consult the requirements of the specific commercial system used before developing a cartographic data base.

The most important attributes in these records, however, are those that locate where the feature should be placed on the map. These attributes are recorded in the form of XY coordinate pairs based upon a Cartesian coordinate system upon which the map is registered when it is digitized. The coordinate system allows the location of each point on the map to be uniquely identified. This unique location identification is critical to automated mapping software—otherwise all points would be displayed on top of each other, as would all lines, polygons, etc. (A more complete description of coordinate systems for geographic information systems is provided in Chapter 6.)

The automated mapping functions of geographic information systems read the values of the various attributes of each record in the cartographic data base and perform manipulations on them that display the features on a map that is in a form familiar to the human eye.

Figure 4.1 is an example of a simple map that has been produced in such a manner. It is a map of a one-block area in "GIS County" (see the exercise at the end of this chapter). The block is bounded by four streets: Beckius Way on the North, Vanderheyden Way on the East, Bertrand Drive on the South, and Winter Court on the West.

The automated mapping functions of a geographic information system need only eight cartographic data-base records to plot the map in Figure 4.1. These eight records include four records to display the street lines and four records to display the text (the names of the streets that the lines represent). Each of these records contains values of the attributes needed by the system to produce the output shown in Figure 4.1.

In order to identify the map locations for these eight features, the map must first be registered on a coordinate system that allows each point to be identified uniquely as XY coordinate pairs. Figure 4.2 shows such a coordinate system superimposed on the map of Figure 4.1. When the map of Figure 4.1 is digitized during the map conversion process, the beginning and ending points of each line and the beginning points of text lines are stored as attribute values in the data-base records of each object on the map. The line representing Beckius Way begins at

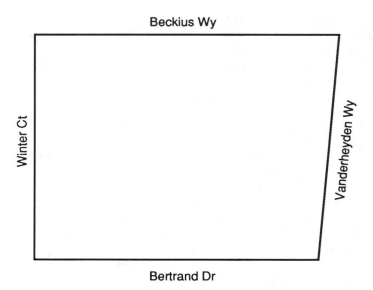

FIGURE 4.1. A simple map of one block bounded by four streets.

FIGURE 4.2. A coordinate system superimposed on the map of Figure 4.1. The coordinates defined by this system locate each line and text string for plotting the map.

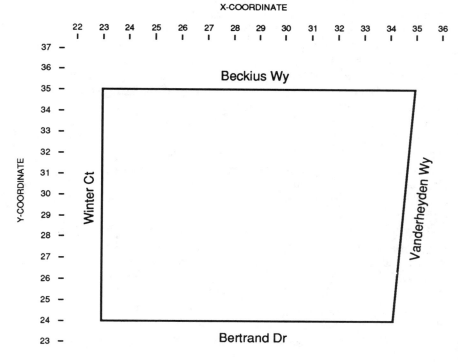

an X coordinate of 23 and a Y coordinate of 35. It ends at X coordinate 35 and Y coordinate 35. The line representing Winter Court begins at X coordinate 23 and Y coordinate 24, represented throughout the remainder of this book as (23,24), where the first value is the X coordinate and the second value is the Y coordinate. The line, then, ends at (23,35). Similarly, Bertrand Drive begins at (23,24) and ends at (34,24), while Vanderheyden Way extends from (34,24) to (35,35). These values, the coordinates of the beginning point and the ending point, must be stored in the data-base record for each line in order for the lines to be plotted by the system. Similarly, the coordinates of the beginning points for the text records must also be recorded in order for the street names to be plotted by the system. The eight cartographic data-base records defining the map of Figure 4.1 are presented in Table 4.2.

The system reads the data stored in records 1 through 4 of Table 4.2 to display the lines shown in Figure 4.1. It reads the attribute values of Record 1 and determines that a "thin," "solid," "yellow," "line" should be plotted starting at coordinate location (23,35) and ending at coordinate location (35,35). Record 2 tells it to plot another thin, solid, yellow, line from (34,24) to (35,35). Records 3 and 4 tell it to plot the remaining lines of Figure 4.1, completing the street lines of the map. These four data-base records, however, have no information to tell the system to plot the names of the streets on the maps. To the system, they are merely lines, representing nothing more than lines. To display the names of the streets, the system must also have Records 5 through 8 of Table 4.2, which contain attributes of the text to be output on the

Table 4.2. A cartographic data base for the map in Figure 4.1. The values in these records define what is to be displayed on the map, how it is to be displayed, and where to display it.

Attribute	Record 1	Record 2	Record 3	Record 4
Cartographic object	Line	Line	Line	Line
Symbology	Solid	Solid	Solid	Solid
Weight	Thin	Thin	Thin	Thin
Color	Yellow	Yellow	Yellow	Yellow
Level	05	05	05	05
Beginning point	(23,35)	(34,24)	(23,24)	(23,24)
Ending point	(35,35)	(35,35)	(34,24)	(23,35)

Attribute	Record 5	Record 6	Record 7	Record 8
Cartographic object	Text	Text	Text	Text
Font type	2	2	2	2
Height (in.)	0.5	0.5	0.5	0.5
Angle	0	80	0	90
Color	Blue	Blue	Blue	Blue
Level	06	06	06	06
Beginning point	(27.3,35.5)	(34.5,27.5)	(27.3,23.5)	(22.5,28.5)
Value	Beckius Wy	Vanderheyden Wy	Bertrand Dr	Winter Ct

map. These "Text" records contain the names of the streets as well as the specifications for the letters and numbers used in plotting the text on the map.

The eight cartographic records of Table 4.2 are used by the automated mapping functions to plot the map shown in Figure 4.1. Additional functions allow the map to be easily changed and manipulated by changing the values of the attributes in the records. For example, the color of the street names can be changed by changing the value of each Color attribute of the four Text records and then replotting the map. The size of the letters and numbers can be increased or decreased by changing the Height values in the Text records. The map can be enlarged (or shrunk) by multiplying the XY coordinates of the Line records by some constant factor. It can be rotated by 90 degrees by adding "90" to the values of the Angle attributes of the Text records and by performing geometric calculations on the XY coordinate values of the Line records. There are many geometric and display functions such as these that automated mapping functions perform on the records of a cartographic data base to change the representation of a map when it is output from the system.

Linking Nongraphics Data to a Cartographic Data Base

The entities represented on maps also have attributes that describe their physical and administrative characteristics in addition to their graphical representation on a map. These attributes normally reside in the administrative or "nongraphics" data bases for use in transaction-based information systems and usually have been in place for many years. Some of the attributes that local governments store in these nongraphics data bases include: the type of pavement on a street, the date it was constructed, the diameter of a water main, the slope of a sewer main, etc. These nongraphics data are different from the cartographic data of map features because they describe the physical entity, while the cartographic data describe the object displayed on the map. Both data bases, however, contain information about the same features.

Nongraphics data consist of attributes of cartographic entities that are needed to describe the physical characteristics of entities in the real world.

Automated mapping and facilities management (AM/FM) systems extend the map manipulation functions of automated mapping systems by allowing nongraphics data to be associated with the cartographic data for display and further manipulation. For example, many cities collect and maintain computerized data about the age and condition of the street pavements in their jurisdictions. The information is used for preparing capital improvement plans for roadway resurfacing or re-

placement and to determine costs and priorities. Water-main break information is often computerized, not only to dispatch repair crews, but
also to monitor trends that help determine water main replacement
priorities. AM/FM systems allow these nongraphics data to be linked
to digital map information so that the data can be analyzed geographically and displayed directly onto a map without digitizing or otherwise
adding it to the map manually.

Nongraphics data bases that contain information about cartographic
features (such as street segments) can be linked to the cartographic
objects on a map by including additional attributes in each record. These
additional attributes relate to labels that uniquely identify the cartographic record and also reside in the record of the nongraphics data
base. Thus, by adding an Identifier attribute (and its XY coordinate location) to the Line records of Table 4.2, information describing each
street can be displayed directly onto the map. Table 4.3 and Figure 4.3
show the records with their unique identifiers and which streets they
represent.

Now, with unique identifiers in each record of the cartographic data
base, any nongraphics data base containing information describing the
objects the records represent can be displayed and manipulated geographically. Take, for example, the pavement history data base represented in Table 4.4.

Record Numbers 68 through 71 in Table 4.4 can be linked to the Line
records (Record Numbers 1 through 4) of the cartographic data base in
Table 4.3 because they have the same Identifiers. (With a relational
data-base management system, the two tables would be "joined" as
described in Chapter 2.) Thus, the data in the two records having the
same Identifier value can be combined, and the attribute data about
the street segment can be associated with the street on the map. Once
that association is made, many graphic manipulations are possible.

Table 4.3. **Cartographic data-base records for the four lines of Figure 4.1,
with unique identifiers and their map locations.**

Attribute	Record 1	Record 2	Record 3	Record 4
Cartographic object	Line	Line	Line	Line
Symbology	Solid	Solid	Solid	Solid
Weight	Thin	Thin	Thin	Thin
Color	Yellow	Yellow	Yellow	Yellow
Level	05	05	05	05
Beginning point	(23,35)	(34,24)	(23,24)	(23,24)
Ending point	(35,35)	(35,35)	(34,24)	(23,35)
Identifier	5423	5424	5425	5426
ID point	(29,34)	(33,29)	(29,25)	(24,29)

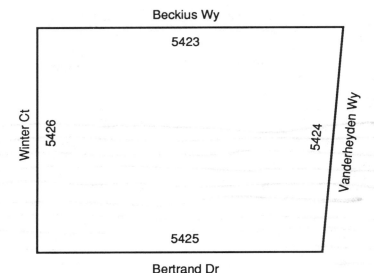

FIGURE 4.3. Map showing the street segments of Figure 4.1 and their unique identifiers.

Display of Nongraphic Attribute Values

A nongraphics data base can be searched on a specific attribute, and the values of that attribute can then be displayed directly on the map. Using the coordinates of the beginning point of the Identifier number, the value of an attribute can be displayed at that location on the map. The attribute Construction Date, for example, can be displayed on the map by placing the value in the Construction Date field of the records in the pavement history data base (Table 4.4) at the XY coordinate given in the ID Point field of the corresponding record in Figure 4.3. Record 68 has a construction date of 1980. Its Identifier, 5423, matches to the

Table 4.4. Five records of a nongraphics data base (pavement history data base). The record for "Beckius Way" (Identifier 5423) describes the pavement for the street segment as "asphalt" pavement type built in 1980 whose current condition is fair.

Data element	Record 67	Record 68	Record 69	Record 70	Record 71
Identifier	5422	5423	5424	5425	5426
Pavement type	Asphalt	Asphalt	Concrete	Concrete	Asphalt
Construction date	1980	1980	1964	1978	1989
Street name	Beckius	Beckius	Vanderheyden	Bertrand	Winter
Street type	Way	Way	Way	Drive	Court
Address range	1100–1199	1010–1099	400–499	1020–1099	400–499
Condition	Fair	Fair	Poor	Good	Excellent

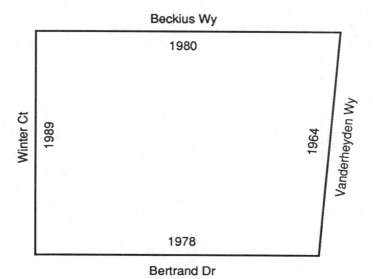

FIGURE 4.4. The display of nongraphics data on the map of Figure 4.1. Here the construction date values of the pavement history data base are displayed at the locations of the unique identifiers of each street segment.

Identifier in Record 1 of Table 4.3, so the value 1980 should be placed at the coordinate of the ID Point of Record 1: (29,34). The same logic is applied to the remaining three records of the pavement history data base to display Construction Date on our map as shown in Figure 4.4.

Display of Selected Nongraphic Attribute Values

By searching through the nongraphics data base for records that have an attribute with a specific value, a map can be produced to display these selected values and the cartographic objects they describe. Figure 4.5, for example, is the result of selecting only those records from the pavement history data base whose Pavement Type attribute contains the value "asphalt." Notice that Records 67, 68, and 71 of Table 4.4 are the only records that have "asphalt" as a Pavement Type. Their Identifiers (5422, 5423, and 5426) are then matched to the Identifiers in Table 4.3, obtaining the XY coordinates of the ID Points so that the text "asphalt" can be displayed on the correct location of the map. (Since Identifier number 5422 is not in Table 4.3, Record 67 cannot be used in this display.) The result of this match is that Record 1 (Identifier 5423) and Record 4 (Identifier 5426) of Table 4.3 represent street segments that have a Pavement Type of "asphalt." A plot of the map showing all street segments constructed of asphalt, then, is given in Figure 4.5.

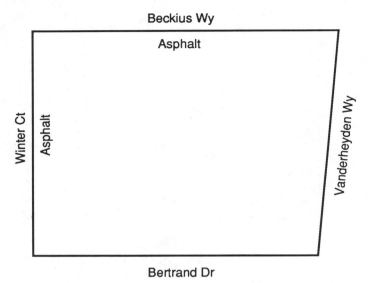

FIGURE 4.5. The display of selected records from a nongraphics data base. Only those street segments with a Pavement Type of "asphalt" have the value "asphalt" displayed at the location of their unique line identifiers.

Changing Map Features Based on Nongraphic Attributes

Based upon the value of a nongraphics attribute, the attribute values of the corresponding cartographic objects (such as the color, line weight, etc.) can be changed to highlight certain information. This means that the street segments on our sample map can be color coded according to the condition of the pavement that is recorded in the pavement history data base. By again matching Identifier values between the records of Tables 4.3 and 4.4, the Color attribute value of the street lines can be changed from all being yellow to a color that is dependent upon the value of the Condition attribute in the pavement history data base. For example, all records which are coded as "poor" condition can have the Color attribute value of their corresponding line record changed from "yellow" to "red." All "excellent" condition records can have their corresponding line record color changed to "blue." "Good" condition records can have their line record color changed to "green," leaving "fair" condition streets as "yellow." The resulting map, output on a color plotter or screen, would have each street segment color coded according to the condition of the pavement of the street.

The same logic can be applied to the thickness of the lines on the map. The Condition values in the pavement history data base can be used to change the Weight values of their corresponding line records in the cartographic data base. In Figure 4.6, the Weight values of the

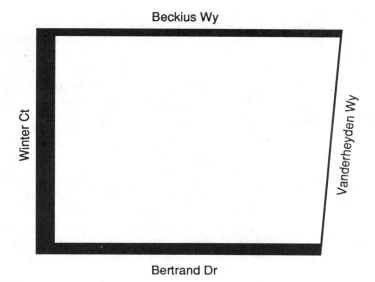

FIGURE 4.6. The result of changing map features based upon the values of attributes from a nongraphics data base. Here the thickness of the lines representing street segments is determined by the values of the Condition attribute of the pavement history data base.

lines have been changed so that those in the best condition are the thickest and those in the worst condition are thinnest.

Adding Topological Data Structures to a Cartographic Data Base

Topological data structures provide additional intelligence to the information stored in a cartographic data base. They tell the computer which cartographic objects are connected to each other logically (i.e., a polygon is bounded by certain lines that have certain beginning points and ending points). Thus, topological data structures define how points, lines, and polygons are related to each other on a map—an implicit relationship that is usually obvious to the human eye, but not explicitly defined to a computer when it reads through the cartographic records as they have been presented so far. Topological data consist of records that contain attributes defining, for example, that a line represents a street segment that connects two intersections (points) and is also one side of a block (polygon). Topological data also define a street intersection (point) as a node on a street network, linking two or more street segments in a continuous path.

Chapter 2 explained that these logical connectivity relationships among cartographic objects (points, lines, and polygons) must be explicitly defined to a geographic information system. Using the map in Figure 4.1

(pg 56)

as an example, a topological data structure can be explicitly defined by three tables in a GIS that contain attributes defining the relationship among the five points, four lines, and single polygon that are related on the map. (Note: the points and lines defined in the preceding sections of this chapter have not defined the topology of the map. They have merely defined the attributes necessary to plot the cartographic objects and to display information about the nongraphics characteristics of the entities represented on the map.)

Point Tables

The point table identifies each point on the map that is related to any line or polygon displayed on the map. Just like the ID Point of Table 4.3, some points are not actually displayed, but rather are used as a reference to some other object displayed on the map (the point that represents the centroid of the block polygon, for example, is the fifth significant point of the map in Figure 4.1 and cannot be seen). The point table contains attributes that not only uniquely identify points for further processing, but also locate them on a map using their XY coordinates and identify other cartographic objects that relate to them.

The five significant points of the map in Figure 4.1 that require explicit definition in a point table include four street intersection points (P1–P4) and one block centroid point (P5). Figure 4.7 shows the locations of these points and their unique identifiers. The point table for these points would contain values such as shown in Table 4.5.

FIGURE 4.7. Five significant points on the map of Figure 4.1. Points P1 through P4 identify street intersections and point P5 identifies the centroid of the block.

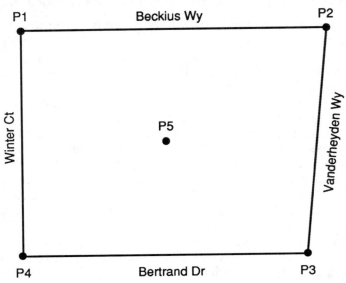

Table 4.5. A topological data structure for the map of Figure 4.1. All significant points, lines, and polygons are uniquely identified and logically linked together by attributes that define their relationships.

Point table

Identifier	P1	P2	P3	P4	P5
Feature	Intersection	Intersection	Intersection	Intersection	Block Centroid
Feature name	Winter Ct and Beckius Wy	Beckius Wy and Vanderheyden Wy	Vanderheyden Wy and Bertrand Dr	Bertrand Dr and Winter Ct	Block #1
XY coordinates	(23,35)	(35,35)	(34,24)	(23,24)	(29,29.5)

Line table

Identifier	5423	5424	5425	5426
Feature	Street segment	Street segment	Street segment	Street segment
Feature name	Beckius Wy	Vanderheyden Wy	Bertrand Dr	Winter Ct
Left polygon	—	—	—	—
Right polygon	1	1	1	1
Beginning point	P1	P2	P3	P4
Ending point	P2	P3	P4	P1

Area table

Identifier	1
Feature	Block
Feature name	Block #1
Centroid	P5
Boundaries	5423, 5424, 5425, 5426

The points identified in a point table form the references that are used for line and polygon definitions. They are used to identify the beginning and ending points of lines, the centroids and vertices of polygons, and locations along a line or inside a polygon. They also can be used as unique references not associated with lines or polygons if such associations are not desired. (A point defining a parcel of land, for example, may not be associated with the property boundaries or the parcel polygon if such association is not necessary. See Chapter 6.)

Line Tables

The line table identifies each line on the map that is related to significant points or polygons. It contains attributes that not only uniquely identify the lines for further processing (as in the pavement history data base example in the previous section), but these attributes also identify what type of feature the line represents and which points and polygons are connected to them.

Obviously, there are four lines in the map of Figure 4.1 that must be defined in a line table: one for each street segment. (Note that these lines represent segments of streets, intersection to intersection, and not the entire length of the streets.) These lines and their identifiers are the same as those shown in Figure 4.3 of the previous section. A line table for Figure 4.3 might contain the attributes listed in Table 4.5.

The line table not only defines each line and what it represents, but it also identifies the relationship each line has with points and polygons on the map. The identifiers of each Beginning Point and Ending Point are stored in the line table, as are the identifiers of each polygon with which the line is associated. (In Figure 4.7 there is only one polygon; so the Left Polygon, starting at the Beginning Point of the line and looking at the Ending Point, has no value, while the Right Polygon side of each line is Block 1.)

Area Tables

The area table identifies all the polygons on the map that might be processed by a geographic information system. It contains attributes that uniquely identify each polygon (for attaching nongraphics data or for spatial analysis applications) and also identify the type of feature it represents and which lines form its boundaries.

The map in Figure 4.1 has only one polygon to identify: the block bounded by the four street segments. Its area table might contain the attributes listed in Table 4.5.

Notice that the area table contains the identifiers of the boundary lines of the polygon as well as its centroid. The centroid is included as an attribute because there are certain spatial analysis functions of a GIS that require only a point to represent a polygon (such as displaying the total number of houses on the block or counting the total number of blocks in the city that have a population greater than a certain number). The line boundaries are important for many functions (such as shading the area of the polygon, counting the number of occurrences—crimes, survey returns, etc.—within the area, or combining polygons of different features—zoning, aldermanic district, etc.

Examples of Using Topological Data Structures

Table 4.5 combines the point, line, and area tables into one topological data structure for the map in Figure 4.1. Since all of the significant features of the map have been identified in the cartographic data base and then related through the topological data structures, the intelligence of the digital map has been significantly enhanced. Not only can the cartographic objects be plotted and linked to other nongraphics data for geographic inquiry and display, but the logical relationships connecting the objects can also be used for a wide range of spatial analysis and inquiry functions.

The data in Table 4.5 can answer the simple geographic question: "Which streets form the boundaries of Block 1?" by searching the area table for the record which has "Block 1" as its Feature Name and finding the line identifiers (Boundaries) that form its boundary. These identifiers are then matched to the Identifiers in the line table to obtain Feature Names for all that match (see Figure 4.8). The result can then be either plotted on a map or listed in a tabular report. This process cannot be achieved by using only the cartographic data in Records 1 through 8 of Table 4.2 because they do not identify Block 1 and because the relationships among the line records are not explicitly defined. What Records 1 through 8 can do is merely provide the system with data that allow it to display or plot the street segments and their names.

The point table and line table can also be used to count how many street intersections there are in Block 1. This procedure involves a search of the point table for all Features that contain the value "intersection." For each one found, the line table is searched for the record that has the same point Identifier in either the Beginning Point or Ending Point attribute of the line. When it is found, the Left Polygon and Right Polygon attributes must be checked for a "1," and a counter is then incremented if it is a "1." When all records of the point table that have an "Intersection" value as a Feature are processed in this manner, the final value of the counter is the number of intersections (four) in Block 1 (see Figure 4.9).

From these two examples, it is clear that the eight cartographic records needed to plot the map in Figure 4.1 (Record numbers 1 through 8 of Table 4.2) do not contain enough data to answer many types of

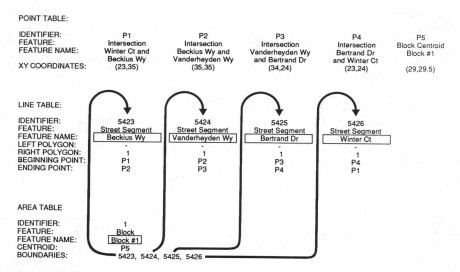

FIGURE 4.8. The use of a topological data structure to find the boundaries of a polygon. The boundary line identifiers (Boundaries) of the area table record for Block 1 point to line table records. The Feature Name of each line table record contains the name of the street segment on the boundary of Block 1.

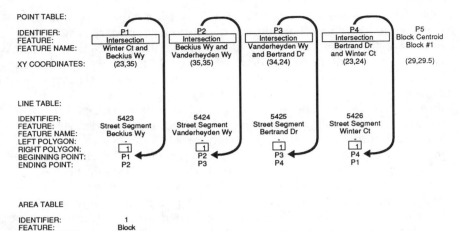

POINT TABLE:

IDENTIFIER:	P1	P2	P3	P4	P5
FEATURE:	Intersection	Intersection	Intersection	Intersection	Block Centroid
FEATURE NAME:	Winter Ct and Beckius Wy	Beckius Wy and Vanderheyden Wy	Vanderheyden Wy and Bertrand Dr	Bertrand Dr and Winter Ct	Block #1
XY COORDINATES:	(23,35)	(35,35)	(34,24)	(23,24)	(29,29.5)

LINE TABLE:

IDENTIFIER:	5423	5424	5425	5426
FEATURE:	Street Segment	Street Segment	Street Segment	Street Segment
FEATURE NAME:	Beckius Wy	Vanderheyden Wy	Bertrand Dr	Winter Ct
LEFT POLYGON:				
RIGHT POLYGON:	1	1	1	1
BEGINNING POINT:	P1	P2	P3	P4
ENDING POINT:	P2	P3	P4	P1

AREA TABLE

IDENTIFIER:	1
FEATURE:	Block
FEATURE NAME:	Block #1
CENTROID:	P5
BOUNDARIES:	5423, 5424, 5425, 5426

FIGURE 4.9. The use of a topological data structure to count the number of points in a polygon. The Identifier of each point table record having a Feature value of "Intersection" is linked to a line table record that has the same identifier in either the Beginning Point or the Ending Point attribute. If the Left Polygon or the Right Polygon attribute of the matching line table record has a value of "1" (for Block 1), then that intersection point is considered in the polygon, Block 1.

geographic questions and perform analysis functions. While they can be used to display the map and nongraphics data describing the features on the map, they are unable to provide information to the geographic information system for the types of geographic questions and analyses described above and shown in Chapter 3.

Topological data are attributes that define the logical connectivity of points, lines, and areas for geographic inquiry and spatial analysis functions of a GIS.

GIS County Exercise 1
Topological Data Structures

Figure E4.1 is a map of GIS County. It is not a large county, but it is very special. It is the only county in the nation that has a county-wide geographic information system to help manage its information.

You, however, must build the topological data structures from this map and the following map showing point and line labels (Figure E4.2) before GIS County officials can use the system. You must build three tables for the system (see samples in Figure E4.3) and answer five geographic questions using these tables:

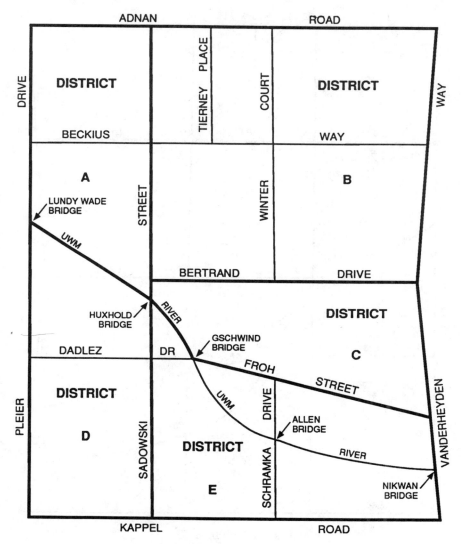

E4.1. General map of GIS County.

1. Point table—List each point, the type of feature it represents, and the name of the feature.

2. Line table—List each line, its beginning and ending points, the polygon (district) on its left side and the polygon on its right side (as though you were standing at the lower numbered point looking at its higher numbered point), the type of feature it represents, and the name of the feature.

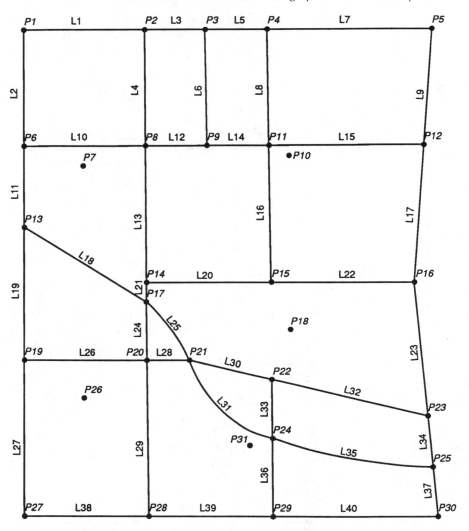

E4.2. Point and line identifiers for GIS County.

3. Area table—List each area (polygon), the point at its centroid, the type of feature the polygon represents, the name of the feature, and each vertex of the polygon.

After completing the tables, answer the following questions using only the tables (do not use the maps). As you answer each question, write down the procedures you used to obtain your answers.

GIS County Exercise 1.2

4. Name the features that are boundaries of each district.

District   *Boundary Features*

5. Which districts have bridges? Include bridges that are on the boundaries of districts.

Bridge   *District(s)*

GIS   County

POINT TABLE

POINT LABEL	FEATURE TYPE	FEATURE NAME
P1	Intersection	Adnan Rd. and Pleier Dr.

LINE TABLE

LINE LABEL	BEGINNING POINT	ENDING POINT	LEFT POLYGON	RIGHT POLYGON	FEATURE TYPE	FEATURE NAME
L1	P1	P2	–	A	Road	Adnan Rd

AREA TABLE

POLYGON LABEL	CENTROID	FEATURE TYPE	FEATURE NAME	VERTICIES
A	P7	District	District A	P1 P2 P17 P13 P1

E4.3. Samples of point table, line table, and area table for GIS County.

6. Name the intersections that are entirely within districts (not on district boundaries).

Intersection   *District*

7. How many intersections are there in each district? (Include both those within and on district boundaries.) Where would you display these numbers if you were to plot a map of total intersections for each district?

District   *Total Intersections*   *Display Point*

8. Hurricane Sarup causes the UWM River to overflow its banks, wiping out all the bridges in the county. At which intersections must barricades be placed in order to reroute traffic?

Bridge   *Intersection*

References

Moellering, Harold (ed.) (1988), *Issues in Digital Cartographic Data Standards (Report #7)*, National Committee for Digital Cartographic Data Standards, The Ohio State University, Columbus, Ohio.

ADDITIONAL READINGS

Corbett, James P. (1979), "Topological Principles in Cartography," Technical Paper #48, U.S. Bureau of the Census, Washington, D.C.

5

Geographic Base Files

The trouble with geography is that there are so many different kinds in local government. The geography of water main valves requires a rather precise reference such as an XY coordinate or distance from a coordinate. The geography of land parcels and buildings requires either points representing their centroids or polygons representing their areas. Sewer mains, street pavements, and easements require a geography that represents line segments, sometimes represented by address ranges and sometimes represented by XY coordinates of beginning and ending points. The geography of political and administrative districts requires line segments (usually street segments or natural features) to identify boundaries or points to identify centroids of polygons. These geographic references (XY coordinates, addresses, address ranges, districts, etc.) are referred to as *geocodes*. Geocodes are the identifiers *(geographic codes)* assigned to map features and nongraphics data-base records that allow data to be processed geographically *(geoprocessing)*.

Geocoding and Geoprocessing

In order to aggregate information vertically for higher-level management and policy functions or to share data horizontally among diverse operational functions, appropriate geocoding capabilities must be available. Since most computerized information systems in local government have been built as single-purpose or transaction-based systems for individual functions or organizational units, the geocodes that they contain are usually unique for each individual unit or function. Thus, building-inspection data, used solely for inspecting buildings for code

147

violations, normally contain geocodes referencing inspection districts, while tax assessment data, used for determining the assessed value of properties, normally contain geocodes referencing tax assessment districts or economic neighborhoods. Rarely are the boundaries of these different districts the same (coterminous). This is because the criteria for establishing the boundaries differ from function to function (inspection workload on the one hand and neighborhood homogeneity on the other). As a result, statistics and summary data shared among different functions cannot be adequately compared or integrated because they represent different geographic areas.

A study of geographic data needs at the City of Milwaukee in the early 1970s revealed a frustrating geocoding dilemma: Despite having current and accurate computerized data, the lack of standard geographic references prevented the expanded use of much of the data. For example, the highest priority of information needed by the Planning Department was accurate and current housing data. In order to distribute limited local and federal funds for improving housing quality to the programs that could produce the most benefits, better descriptive information was needed on the existing housing supply and the current and future demands for housing. The most pressing information need in the Planning Department, then, was for housing characteristics within the city boundaries (see Figure 5.1).

The most accurate and current information on housing characteristics resided on the computerized files of the tax assessment system. As a city function, the property tax assessment system was being updated regularly as assessors obtained information about changes to properties (from building permits, demolition orders, deed changes, inspections, etc.), which caused changes to the property attributes stored on the city's Real Estate Master File. Thus, through the tax assessment function of the city, it was possible to obtain such housing characteristics as: the number of housing units, the assessed value, owner occupancy, construction type, number of bedrooms, and others. Because the data were collected for tax assessment purposes, however, summary statistics (average assessed value, owner-occupancy rate, total multifamily dwelling units, etc.) were available only by tax assessment district (see Figure 5.2).

A significant factor in determining housing need is the demographic character of geographic areas. Data such as female-headed households, number of children under 18 years of age, average income, etc. were available from the computerized data files of the U.S. Census Bureau; however, census geography was different from tax assessment geography. These data were available only by census tract (see Figure 5.3).

Both sources of information (tax assessment records and census data) provided accurate and comprehensive data for the primary information needs of the Planning Department. The use of the data, however, was

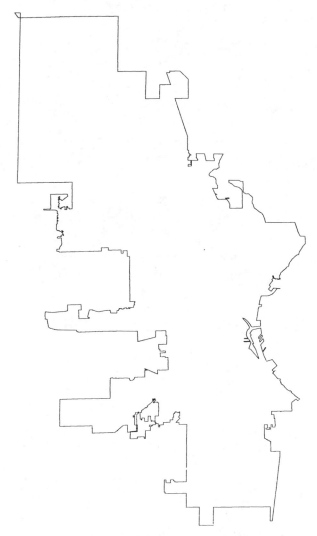

FIGURE 5.1. City outline defining the geographic area for which housing characteristic data is needed.

limited by the dilemma caused by different geographic areas that were not coterminous. Figure 5.4 shows this dilemma by overlaying tax assessment districts and census tracts onto the outline of the city.

Because the boundaries of the two different areas were not coterminous, it was impossible to combine summary statistics from the two sets of data for meaningful housing information. While it was possible, for example, to obtain statistics such as the average value of a housing

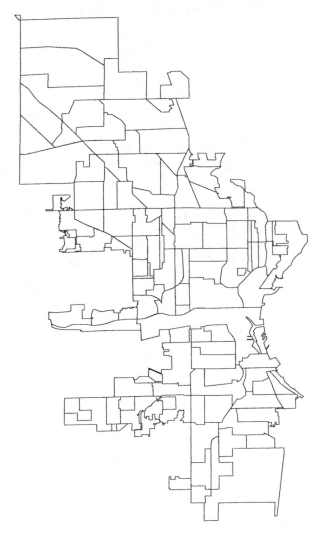

FIGURE 5.2. Tax assessment districts used to code housing data geographically by the city tax assessor.

unit for each tax assessment district and the average household income for each census tract, it was impossible to combine these statistics to determine the average household income as a percent of housing unit value because the geography of the two statistics was different.

Compounding this frustration with different geographic boundaries was the Planning Department's inability to provide housing statistics to the elected officials because aldermanic districts had yet a different set of boundaries (see Figure 5.5). These aldermanic district boundaries were set based upon criteria that were different from the criteria used

to establish tax assessment districts and census tracts—they were established in order to balance population among political representatives. Because these elected officials would ultimately set housing policies for the city, they needed to know the effects of proposed policies
on their constituents prior to voting on them. The summary data they
needed could not be produced by aldermanic district because they were
geocoded to areas which had different boundaries. Figure 5.6 displays

FIGURE 5.3. Census tracts used to code demographic data geographically by
the Planning Department.

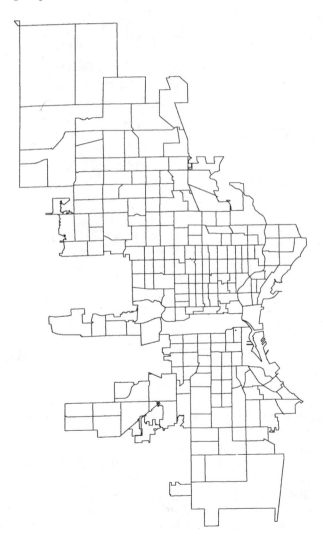

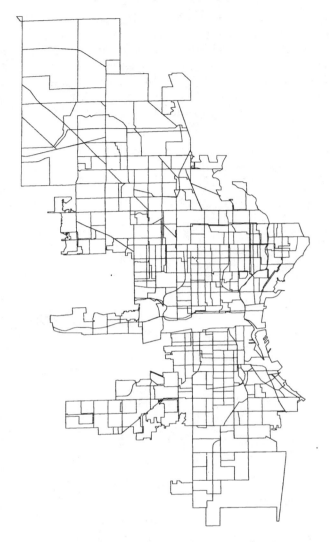

FIGURE 5.4. An overlay of tax assessment districts and census tracts, showing that their boundaries are not coterminous.

aldermanic districts overlaid onto the tax assessment districts and census tracts of Figure 5.4.

Further analysis of geographic information needs citywide revealed that almost every function in city government used geographic areas that were unique only to those functions. There were housing code inspection districts, electrical inspection districts, plumbing inspection districts, public health nursing districts, garbage collection routes, water pressure districts, police squad areas, urban forestry districts, fire re-

sponse zones, zoning districts, and many, many more. In all, there
were over 200 different geographic areas defined for the various management, administrative, and political activities of city government—
and none shared the same boundaries. If all of these areas were overlaid onto one map as the three areas shown in Figure 5.6, the map
might be completely black! (No wonder it was difficult to share data
among different functions of the city.)

What these 200 different geographic areas did have in common,

FIGURE 5.5. Aldermanic districts used by elected officials to assess the impact
of new programs and policies.

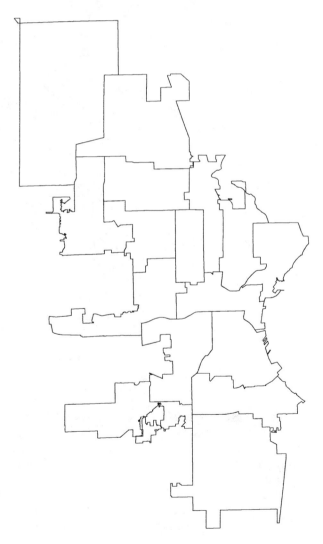

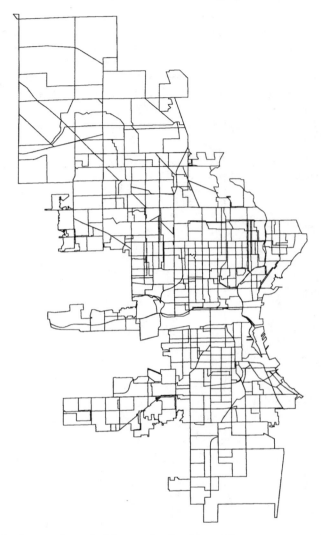

FIGURE 5.6. An overlay of aldermanic districts with tax assessment districts and census tracts.

however, were addresses. Because local government has such a direct impact on individuals and their properties, most of its functions require an address as part of their information base, and these addresses can be located in each of the 200 different geographic areas. A given address, for example, could be located in Census Tract 43, Aldermanic District 9, Housing Code Inspection District 213, Solid Waste Collection Route 3A, Police Squad Area C23, and so on. Conceivably, then, each address could have over 200 geocodes assigned to it in order to be able to summarize data by any geographic area. Figure 5.7 displays the location of every property in the city (over 160,000 dots) that could be assigned a different geocode for each function in the city.

The assignment of geocodes to records in nongraphics data bases is called *geocoding*. Geocoding is a process that is performed either manually or with a computer to assign location identifiers to data for further *geoprocessing* of the data for analysis. (We all have manually geocoded data when we have used a ZIP Code Directory to find a ZIP

FIGURE 5.7. Dot density map of 160,000 parcel centroids. Each dot can be placed within any geographic area used by local government.

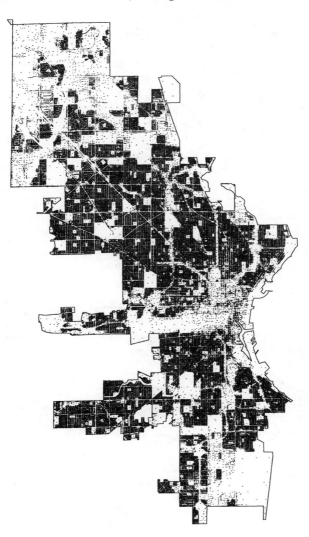

Code for an address.) To assign manually every address in the local government with a geocode for every different type of district and for every different data base used in local government functions, however, is a task no sane human would even consider undertaking. Fortunately, there are computerized methods to perform the geocoding task. Computerized geocoding uses computer programs and a *Geographic Base File* (GBF) to eliminate the time-consuming process of manually geocoding data.

A Geographic Base File is a computerized data file (usually a flat file) that contains only geographic attributes (addresses, census tracts, XY coordinates, administrative districts, political districts, and other identifiers that can be assigned to geographic features). Computer programs read an address in a nongraphics data base and find the record in the GBF that defines the geography in the city for that address. These programs then extract the desired geocodes from that record of the GBF and move them to the record of the nongraphics data base, making them attributes of the features in the nongraphics data (see Figure 5.8). Further processing of the data in the nongraphics data base is then performed using these geocodes as attributes.

Take, for example, a nongraphics data base of tax assessment records that have no geocodes except for the address of each property. In order to use the data base for producing summary statistics such as the average assessed value of properties by census tract, it is necessary first to geocode the data base by census tract. Then a simple program can sort the data on census tract and compute the average assessed value.

FIGURE 5.8. The computerized geocoding process. A geographic base file is used by geocoding programs that read a nongraphics data file and add geocodes to the records in the file.

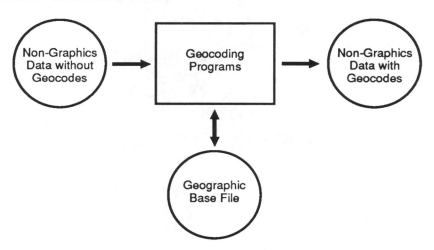

The geocoding process is accomplished by a series of programs that read each record of the tax assessment data base and use the property address in the record to find the appropriate record on the geographic base file. When they find the correct GBF record, they read its census tract attribute and write it to a data field on the tax assessment data-base record. When this is done for all the records in the data base, then the data base has been geocoded, and other programs can then be used to perform manipulations and calculations on the data. This can be done because the tax assessment data base has been expanded by the addition of a census tract attribute to each record in the data base (see Table 5.1).

Elements of a Geographic Base File

Because a geographic base file contains geographic attributes, it is used as a computerized reference file for translating data from one geographic reference to another. It contains two basic sections: one defining street segments and another defining geographic areas. Both sections are based upon traditional methods adopted for defining addresses in municipal government.

Street Segments

Street segments are defined in a GBF by the range of addresses that run along a street from one intersection to the next intersection. The address range is recorded by identifying the lowest possible house number (at one intersection) and the highest possible house number (at the other intersection). All addresses along that street segment are either greater than the lowest house number or less than the highest house number, and they all have the same street name. With no other line intersecting the street segment, it becomes the basic geographic unit upon which all other geographic references are based. In Figure 5.9, the street feature, Bertrand Dr., is defined by two street segments: one with an address range of 1020–1099 (from its intersection with Vanderheyden Wy. to its intersection with Winter Ct.), and one with an address range of 1100–1199 (from Winter Ct. to Sadowski St.).

Just like the line segments of the line tables defined for topological data structures in Chapter 4, street segments in a GBF are identified by a label that is unique for the entire geographic area. In fact, the lines in a topological data structure that also are street segments in a GBF have the same labels because they represent the same map features. Table 5.2 gives the attributes of the street segment portion of a GBF with unique segment labels.

Table 5.1. Tax assessment data (a) before geocoding and (b) after geocoding.

Parcel No.	Address	Assessed Value
047	428 N. Sadowski St	$13,520
048	484 N. Winter Ct	13,520
049	460 N. Winter Ct	13,520
050	446 N. Winter Ct	13,520
051	420 N. Winter Ct	13,520
052	412 N. Winter Ct	13,520
053	410 N. Winter Ct	15,430
054	406 N. Winter Ct	14,750
055	419 N. Vanderheyden Wy	87,670
056	1125 W. Bertrand Dr	76,430
057	1119 W. Bertrand Dr	76,430
058	1113 W. Bertrand Dr	76,430
059	1107 W. Bertrand Dr	79,500
060	1101 W. Bertrand Dr	79,500
061	1099 W. Bertrand Dr	79,500
062	1087 W. Bertrand Dr	66,800
063	1075 W. Bertrand Dr	66,800
064	1051 W. Bertrand Dr	80,100
065	1033 W. Bertrand Dr	81,000
066	343 N. Vanderheyden Wy	35,000

(a)

Parcel No.	Address	Assessed Value	Census Tract
047	428 N. Sadowski St	$13,520	34
048	484 N. Winter Ct	13,520	34
049	460 N. Winter Ct	13,520	34
050	446 N. Winter Ct	13,520	34
051	420 N. Winter Ct	13,520	34
052	412 N. Winter Ct	13,520	34
053	410 N. Winter Ct	15,430	34
054	406 N. Winter Ct	14,750	34
055	419 N. Vanderheyden Wy	87,670	34
056	1125 W. Bertrand Dr	76,430	52
057	1119 W. Bertrand Dr	76,430	52
058	1113 W. Bertrand Dr	76,430	52
059	1107 W. Bertrand Dr	79,500	52
060	1101 W. Bertrand Dr	79,500	52
061	1099 W. Bertrand Dr	79,500	52
062	1087 W. Bertrand Dr	66,800	52
063	1075 W. Bertrand Dr	66,800	52
064	1051 W. Bertrand Dr	80,100	52
065	1033 W. Bertrand Dr	81,000	52
066	343 N. Vanderheyden Wy	35,000	52

(b)

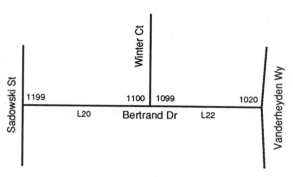

FIGURE 5.9. Two segments of a street showing their address ranges from intersection to intersection. All properties with addresses between 1020 and 1099 on Bertrand Dr. are located on the Bertrand Dr. street segment from Vanderheyden Wy. to Winter Ct. (Segment L22). Those with addresses between 1100 and 1199 are located on the Bertrand Dr. segment from Winter Ct. to Sadowski St. (Segment L20).

Geographic Areas

The geographic areas represented in a GBF are defined by the street segments that form their boundaries. Census tracts, for example, are geographic areas defined by the U.S. Census Bureau for statistical reporting of demographic, economic, and other data. The boundaries of census tracts are either major streets, natural features (such as rivers, lakes, etc.), or corporate limits of the local government. (In some cases, other geographic features may form the boundaries of census tracts. Examples include railroads, freeways, and other manmade features.) In addition, the boundaries of each census tract were established so that the population in each was approximately 4000 people, regardless of its size. Census tracts are numbered consecutively and are unique within the jurisdiction.

Many applications require smaller geographic areas than census tracts. To accommodate this need, the Census Bureau provided for a census tract subdivision scheme that is smaller than a census tract, yet can be aggregated into a census tract when necessary. The Census Bureau di-

Table 5.2. The line segment portion of a geographic base file for the street segments shown in Figure 5.9. The address 1148 Bertrand Dr. can be assigned street segment L20 while the address 1097 Bertrand Dr. can be assigned street segment L22.

Segment label	House number (low)	House number (high)	Street name
L20	1100	1199	Bertrand Dr.
L22	1020	1099	Bertrand Dr.

vided each census tract into block groups, which comprise a population of approximately 1000 people and which have boundaries that are
coterminous with census tract boundaries.

Block groups, in turn, are built from individual blocks, which are the
smallest geographic areas defined by the Census Bureau for geocoding
data. This is because blocks are the smallest polygons that can be formed
by street segments. They are uniquely identified within a census tract
by a three-digit numeric code, the first position of which represents the
block group and the last two positions of which are consecutively numbered so that they are unique within the block group. Block 101, then,
is the first block numbered within Block Group 1 and Block 304 is the
fourth block numbered within Block Group 3. By appending the census
tract number to the front of the block number, each block can be uniquely
defined across the entire jurisdiction (there is only one Block 304 in
Census Tract 1008, although there may be a Block 304 in Census Tract
892).

To help local jurisdictions use this geographic coding scheme, the
Census Bureau provides a series of maps that identify each block number, block group, and census tract in all areas of the nation with populations greater than 50,000. This Metropolitan Map Series is a set of
maps that were created for the 1970 census, covering the urbanized
portion of every Standard Metropolitan Statistical Area (SMSA) in the
country. Figure 5.10 shows a portion of a Metropolitan Map Series map
with block numbers for Census Tract 1008.

Local geographic areas other than census tracts are defined by each
jurisdiction without assistance from the Census Bureau. They are similar to census tracts in that they are defined by boundaries that are
streets, natural features, corporate limits, railroads, or freeways, yet
they differ from census tracts in the criteria used to establish their size.
Some local geographic areas are based upon a larger population (aldermanic district). Some are based upon area (quarter-sections) or travel
time (fire and emergency medical service response zones). Others are
based upon demographic characteristics (neighborhood), economic
characteristics (tax assessment district), historic character (historic pres-
ervation districts), or legislation (tax incremental districts). In many cases,
however, the size of local geographic areas is based upon the amount
of work required to deliver public services (solid waste collections,
building inspections, property assessments, calls for police service, water
meters to read, etc.). Figures 5.2 and 5.5 are two examples of local
geographic areas.

A geographic base file defines the geographic areas of a jurisdiction
in terms of their relationship to street segments (see Figure 5.11). Since
the boundaries of census tracts, blocks, and most all local geographic
areas coincide with street segments, the geographic areas are ap-
pended to the street segment portion of the GBF records. Street segments, however, usually have two sides (odd-numbered and even-

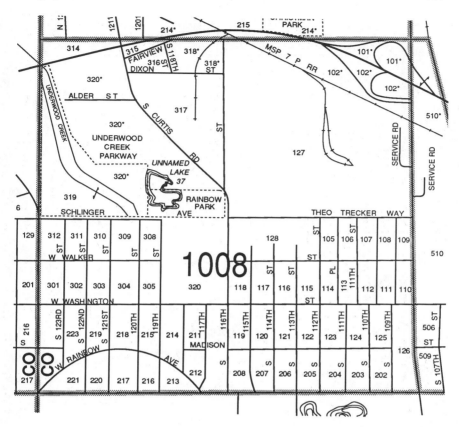

FIGURE 5.10. Block numbers for Census Tract 1008 from a Metropolitan Map Series map.

numbered addresses) associated with different blocks and, possibly, different census tracts and different local geographic areas. In order to relate geographic areas to street segments, then, a GBF must differentiate between the "even" side of the segment and the "odd" side of the segment. This is accomplished either by defining separate records (one for the even-side addresses and one for the odd-side addresses) or by defining even-side and odd-side geographic areas within the street segment record itself (see Table 5.3).

Many network-based geographic features do not relate directly to street segments and thus are more difficult to geocode using a street-segment-based geographic base file. A water main, for example, may run down a street and then change direction in the middle of the segment. A sewer main may change slope in the middle of a street segment. Many street segments may contain more than one segment of water main or sewer main because of differences in size, year of construction, type of

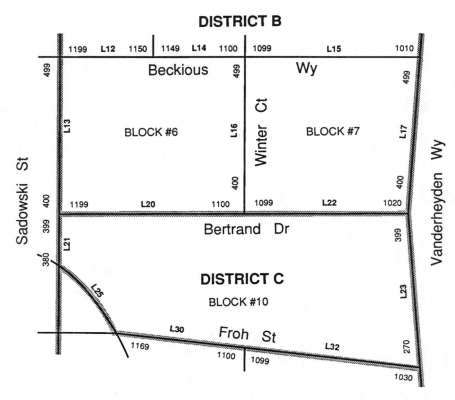

FIGURE 5.11. Address ranges and geographic areas for street segments surrounding three blocks.

Table 5.3. A geographic base file for the map shown in Figure 5.11. Each street segment record contains "even-side" and "odd-side" geocodes, depending on the parity of the address on the street segment.

Segment label	House number (low)	House number (high)	Street name	Even-side block	Odd-side block	Even-side district	Odd-side district
L12	1150	1199	Beckius Wy.	—	6	B	B
L13	400	499	Sadowski St.	6	—	B	—
L14	1100	1149	Beckius Wy.	—	6	B	B
L15	1010	1099	Beckius Wy.	—	7	B	B
L16	400	499	Winter Ct.	7	6	B	B
L17	400	499	Vanderheyden Wy.	—	7	—	B
L20	1100	1199	Bertrand Dr.	6	10	B	C
L21	380	399	Sadowski St.	10	—	C	—
L22	1020	1099	Bertrand Dr.	7	10	B	C
L23	270	399	Vanderheyden Wy.	—	10	—	C
L30	1100	1169	Froh St.	10	—	C	—
L32	1030	1099	Froh St.	10	—	C	—

162

material, or other characteristic. Similarly, a street pavement surface may span one street segment, cross a street intersection, and end in the middle of the next segment, where another surface with different characteristics (year installed, type of surface, etc.) begins. Depending upon specific needs, which may be unique to a local government, some modification to a street-segment-based GBF may be necessary for geocoding these types of features. One method for addressing the problem is called *dynamic segmentation,* which is the establishment of a series of cross-reference tables that link each type of segment to the appropriate street segment(s). Fletcher (1989) refers to these tables as an "indexed chain" because they maintain "explicit point–chain relationships" between points (nodes) on the street segment and the chains (segments) of a nonstreet network.

Property boundaries, easements, and zoning boundaries create a different problem in using a GBF. These geographic features, being polygon based rather than segment based, are areas that are contained within the smallest geographic areas of a GBF (blocks), and therefore their boundaries do not always coincide with street segments. The street-segment-based GBF cannot be used to geocode these features unless they are referenced only by an address and not by their shape. Chapter 6 discusses the geographic referencing of property boundaries and other small area features in a geographic information system.

Types of Geographic Base Files

DIME Files

One of the most popular geographic base files used by many local governments across the nation is the GBF/DIME File, which was created by the U.S. Bureau of the Census in order to conduct the 1970 Census of Population and Housing by mail for the urbanized portion of the U.S. While its predecessor, the Address Coding Guide (ACG), had been developed in conjunction with the Metropolitan Map Series to inventory all address ranges in urban areas and relate them to Census Bureau geography (census tracts, block groups, and blocks), the DIME File extended this type of information by providing the foundation for automated mapping of the street segments and other segment-based geographic features. Thus, the DIME File was created not only to geocode address-based data (as described in the previous section), but also to be used in a geographic information system for automated mapping functions.

DIME is an acronym for dual independent map encoding, which refers to the combination of two separate sets of information—the *nodes* at the endpoints of line segments and the *areas* enclosed by the line segments—into one computer file. Representing the areas in terms of their relationship to line segments has been described in the previous section. Representing the nodes and their relationship to line segments is a matter of identifying the XY coordinates of the endpoints of each line segment and labeling each with a unique identification number.

Figure 5.9, for example, displays two line segments representing a street. Each segment, L20 and L22, has a beginning point and an ending point. By recording the XY coordinate of each point (node), a straight line connecting them can be drawn to represent the street segment. When two line segments intersect or otherwise connect (as the two segments of Figure 5.9), the XY coordinates of the beginning node of one segment are the same as the XY coordinates of the ending node of the other segment. Thus, two or more connected line segments share one node in common. When the nodes are labeled with unique identifiers and a coordinate system is used to register their locations, the street in Figure 5.9 is represented as shown in Figure 5.12. Since the street, Bertrand Dr., consists of two street segments (L20 and L22) and is a continuous street spanning two blocks, the two segments share one node in common (P15). The other nodes (P14 and P16) are shared with street segments on Sadowski St. and Vanderheyden Wy.

The DIME File is structured such that each record contains both even-side and odd-side address geography. In DIME terminology, this reference is called "from" and "to" and "left" and "right." Thus, a "from node" is the segment endpoint that is at the low end of the address range along the street and the "to node" is at the high end of the address range. If one were standing at the "from node" facing the "to node" (as addresses increase), the geographic areas to the left are labeled "left" and those to the right are labeled "right." Thus, the DIME File contains "from nodes," "to nodes," "left address ranges," "right address ranges," "left" geocodes, and "right" geocodes, as shown in Table 5.4.

FIGURE 5.12. Nodes and coordinates for the street segments of Bertrand Dr. The nodes form the endpoints of the street segments and are located on a map by their XY coordinates.

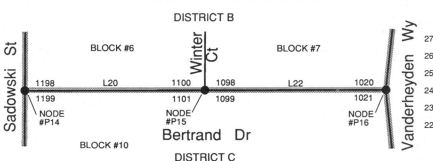

Table 5.4. DIME File information for the two street segments of Figure 5.12.

Segment ID	L20	L22
From node	P15	P16
From node coordinates	(23,24)	(34,24)
To node	P14	P15
To node coordinates	(12,24)	(23,24)
Street name	Bertrand Dr.	Bertrand Dr.
Left address (low)	1101	1021
Left address (high)	1199	1099
Right address (low)	1100	1020
Right address (high)	1198	1098
Left block number	10	10
Right block number	6	7
Left district code	C	C
Right district code	B	B

The topological structure of the DIME File allows an editing capability to ensure that the data in the file are accurate and that all land in the jurisdiction is represented. By selecting all records in the DIME File that have the same block number in either the Left Block Number field or the Right Block Number field, for example, the node numbers of each segment can be checked to ensure that the segments are "chained" together around the perimeter of the block. They are completely chained together when the "from node" of one segment is the same as the "to node" of another and the "from node" of that segment is the same as the "to node" of a third segment, and so on for the entire block. Similarly, the boundaries of other geographic areas can be checked to ensure that they are chained together by checking only those segments having different values in the Left-Side field and Right-Side field for a given geographic area (this eliminates all segments that are entirely within the area). If the "from nodes" and "to nodes" of these boundary records are chained together as described for blocks, then the geographic area is completely enclosed by street segments. Segments that are on the boundary of the entire geographic area represented by the DIME File contain no geographic data on one side of the segment (either the Right-Side geography or the Left-Side geography). Such records are coded with a value of "1" in a field called Coding Limit Flag since the missing geographic codes are not omissions—they are outside the limits of the coding area. All other segments within the coding area have blanks in the Coding Limit Flag field.

The cartographic accuracy of the DIME File can be ensured by checking the XY coordinates within segments having the same node number. If the "from node" of one segment is the same as the "to node" of another segment and their XY coordinates are different, then either the node numbers are wrong or their coordinate locations are wrong. More technical information about the editing capabilities of the DIME

File is available from the Data User Services Division of the U.S. Bureau of the Census (Bureau of the Census, 1980).

Since the DIME File can also be used to produce maps of the jurisdiction, it contains records representing segments that are not street segments, yet do form boundaries of geographic areas or are of such significance as to require display on a map. Such "nonstreet features" include railroads, rivers, shorelines, park boundaries, street extensions, addressed walkways, and political, jurisdictional, or other geographic boundaries that do not conform to physical features. While these records contain the same geographic references as street segments, including XY coordinates of the nodes that form their endpoints, they usually have no address information, since the segments may not provide public access to property. In a DIME File record, these segments are identified by a numeric or alphabetic code in a field called Nonstreet Feature Code.

The DIME File contains two different sets of coordinates for the locations of the nodes: latitude and longitude coordinates and state plane coordinates. These coordinate systems are discussed in detail in Chapter 6.

With the basic segment definitions and coordinate values established in the DIME File, other geographic areas beyond census tracts and blocks can be defined. The Census Bureau has defined some of these areas (ZIP Code, county, etc.), and local jurisdictions have defined others that satisfy specific local needs (building inspection districts, police squad areas, etc.). A detailed list of all elements contained in a DIME File is given in Table 5.5.

As a geocoding tool and as a digital representation of a map, the DIME File has almost unlimited local application possibilities, from geocoding address-based data to reapportioning political districts to combining various local data files for mapping comparison statistics among geographic areas for planning and policy analysis. To be effective, however, the local government must ensure its accuracy, extend its geographic codes to satisfy specific local needs, continually update and maintain the data, and modify its structure to fit local geography. A portion of selected fields from a locally maintained DIME File is given in Table 5.6.

TIGER Files

Inconsistencies between the DIME Files and the Metropolitan Map Series and other geographic data sources during the 1980 census prompted the Census Bureau to develop a new method for preparing a geographic base file and census maps for enumerators for the 1990 census. Because the creation the 32,000 maps and 278 DIME Files in 1980 was accomplished through independent, labor-intensive, and error-prone manual procedures, the Census Bureau decided to combine the map

Table 5.5. A list of elements in a DIME File. The elements listed under "Local geography" are added by local governments as needed.

Record identification number	State Code (right)
"From node"	County Code (left)
"From node" coordinates	County Code (right)
Latitude, longitude	MCD/CCD Code (left)
State plane coordinates	MCD/CCD Code (right)
State plane code	Local geography
"To node"	U.S. Congressional district (left)
"To node" coordinates	U.S. Congressional district (right)
Latitude, longitude	State Senate district (left)
State plane coordinates	State Senate district (right)
State plane code	State Representative district (left)
Segment identification	State Representative district (right)
Street prefix direction	County supervisory district (left)
Street or nonstreet feature name	County supervisory district (right)
Street type	Municipal aldermanic district (left)
Street suffix direction	Municipal aldermanic district (right)
Nonstreet feature code	Tax assessment district (left)
Coding limit flag	Tax assessment district (right)
"From map" (basic and suffix)	Building inspection district (left)
"To map" (basic and suffix)	Building inspection district (right)
Address range	Public Health nursing district (left)
Left address (low)	Public Health nursing district (right)
Left address (high)	Police district (left)
Right address (low)	Police district (right)
Right address (high)	Fire response zone (left)
Census Bureau geography	Fire response zone (right)
Census tract (left)	Emergency medical service area (left)
Census tract (right)	Emergency medical service area (right)
Block number (left)	Water pressure district (left)
Block number (right)	Water pressure district (right)
ZIP Code (left)	Solid waste collection district (left)
ZIP Code (right)	Solid waste collection district (right)
SMSA Code (left)	Water meter route (left)
SMSA Code (right)	Water meter route (right)
Place Code (left)	Neighborhood (left)
Place Code (right)	Neighborhood (right)
State Code (left)	Planning district (left)
	Planning district (right)

preparation process with the geographic base file creation into one integrated system for 1990. In addition, the system expanded the geographic coverage of the geographic base file to include line segments for the entire nation (a total of more than 50 million line segments!). This system is called the *TIGER system*, and the files that create the maps and contain the geographic base information are called the *TIGER Files*.

The term *TIGER* is an acronym for topologically integrated geographic encoding and referencing. TIGER Files contain *topological* data

Table 5.6. One page of a DIME File report listing selected elements from a locally maintained DIME File.

Street name	LLOW ADDR	LHIGH ADDR	RLOW ADDR	RHIGH ADDR	RECORD NUMBER	LEFT TRACT	LEFT BLK	RIGHT TRACT	RIGHT BLK	LEFT ZIP	RIGHT ZIP	LEFT QTR	RGHT QTR	LEFT ALD DIST	RGHT ALD DIST
W Casper St	7401	7499	7400	7498	015806	006	909	006	909	53223	53223	141	141	09	09
W Casper St	7701	7799	7700	7798	015807	006	106	006	111	53223	53223	142	142	15	15
W Casper St	8001	8099	8000	8098	015808	006	208	006	209	53223	53223	142	142	15	15
W Casper St	8101	8199	8100	8198	015809	006	210	006	209	53223	53223	142	142	15	15
W Casper St	8201	8299	8200	8298	015810	006	210	006	209	53223	53223	142	142	15	15
N Cass St	601	627	600	626	015823	153	103	153	120	53202	53202	396	396	04	04
N Cass St	629	699	628	698	015824	153	103	153	120	53202	53202	396	396	04	04
N Cass St	701	749	700	748	015811	143	301	143	209	53202	53202	393	393	04	04
N Cass St	751	799	750	798	015812	143	302	143	206	53202	53202	393	393	04	04
N Cass St	801	899	800	898	015813	143	303	143	204	53202	53202	393	393	04	04
N Cass St	901	933	900	932	015814	143	303	143	204	53202	53202	393	393	04	04
N Cass St	935	999	934	998	015825	143	305	143	306	53202	53202	393	393	04	04
N Cass St	1001	1199	1000	1198	015815	143	308	143	307	53202	53202	393	393	04	04
N Cass St	1201	1299	1200	1298	015816	111	207	111	208	53202	53202	360	360	04	04
N Cass St	1301	1399	1300	1398	015817	111	206	111	205	53202	53202	360	360	04	04
N Cass St	1401	1499	1400	1498	015818	111	201	111	202	53202	53202	360	360	04	04
N Cass St	1501	1599	1500	1598	015819	111	108	111	107	53202	53202	360	360	03	03
N Cass St	1601	1649	1600	1648	015820	111	103	111	104	53202	53202	360	360	03	03
N Cass St	1651	1699	1650	1698	015821	111	103	111	102	53202	53202	360	360	03	03
N Cass St	1701	1749	1700	1748	015826	112	301	112	302	53202	53202	354	354	03	03
N Cass St	1751	1799	1750	1798	015827	112	208	112	303	53202	53202	354	354	03	03
W Cawker Pl	3001	3099	3000	3098	015859	063	207	063	208	53210	53210	309	309	10	10

Street															
N Celina Ct	7701	7799	7700	7798	901501	002	908	002	908	53224	53224	078	078	15	15
N Celina St	8001	8099	8000	8098	015959	002	106	002	105	53224	53224	073	073	15	15
N Celina St	8101	8199	8100	8198	015960	002	106	002	105	53224	53224	073	073	15	15
N Celina St	8201	8299	8200	8298	051767	002	118	002	115	53224	53224	073	073	15	15
N Celina St	8301	8375	8300	8376	900543	002	118	002	115	53224	53224	073	073	15	15
N Celina St	8379	8399	8378	8398	900544	002	118	002	115	53224	53224	073	073	15	15
E Center St	100	198	101	199	016136	082	106	082	201	53212	53212	313	322	06	06
E Center St	200	248	201	225	016137	081	105	081	204	53212	53212	313	322	06	06
E Center St	250	298	227	299	016138	081	105	081	203	53212	53212	313	322	06	06
E Center St	300	398	301	399	016139	081	106	081	202	53212	53212	314	321	06	06
E Center St	400	498	401	499	016140	081	107	081	201	53212	53212	314	321	06	06
E Center St	500	598	501	599	016141	080	105	080	204	53212	53212	314	321	06	06
E Center St	600	698	601	699	016142	080	106	080	203	53212	53212	314	321	06	06
E Center St	700	798	701	799	016143	080	107	080	202	53212	53212	314	321	03	03
E Center St	800	898	801	899	016144	080	108	080	201	53212	53212	314	321	03	03
E Center St	900	998	901	999	016145	079	206	079	203	53212	53212	314	321	03	03
E Center St	1000	1098	1001	1099	016146	079	205	079	204	53212	53212	314	321	03	03
E Center St	1100	1198	1101	1199	016147	079	101	079	101	53212	53212	315	315	03	03
W Center St	101	199	100	198	016150	082	202	082	105	53212	53212	322	313	06	06
W Center St	201	299	200	298	016153	082	203	082	104	53212	53212	322	313	06	06
W Center St	301	399	300	398	016154	083	206	083	203	53212	53212	322	313	06	06
W Center St	401	499	400	498	016157	083	205	083	204	53212	53212	322	313	06	06
W Center St	501	523	500	522	016158	083	304	083	106	53212	53212	323	312	06	06
W Center St	525	599	524	598	016159	083	304	083	106	53212	53212	323	312	06	06
W Center St	601	699	600	698	016160	083	305	083	105	53212	53212	323	312	06	06
W Center St	701	749	700	748	016288	084	306	084	101	53206	53206	323	312	06	06

structures describing how points and lines relate to each other on a map to define geographic areas. The files combine three sources of data (line-segment-based maps produced by the U.S. Geological Survey, existing DIME Files, and geographic area relationship files used for tabulating the 1980 census data) into one *integrated* relational data base. They contain *geographic* feature *references* that can be *encoded* directly onto computer-produced maps (no longer are census maps and census geographic base files created and maintained by separate methods).

While the DIME Files were originally encoded and later maintained from the manually created Metropolitan Map Series maps and covered only about 2 percent of the land area of the nation, the base map information for the TIGER Files was created by the U.S. Geological Survey (USGS) through a raster scanning process of their 1:100,000 scale National Map Series to capture line information for the entire country: roads, hydrography, railroads, and other transportation networks. After vectorizing the scanned lines, this map information was stored within the TIGER Files so that updates and other changes to the map and the geographic references are centralized. This eliminates the inconsistencies between maps and DIME Files that were experienced during the 1980 census.

Unlike the DIME Files, which utilize a flat file structure containing point, line, and area information entirely within each record, the TIGER Files are relational in structure, with all geometric and topological relationships and feature attributes stored either explicitly in the tables of the data base, or implicitly in the data-base structure itself (Broome, 1984). A relational structure was chosen, according to Broome, because:

> The planned many uses of the TIGER file preclude it (from) being hierarchically structured as are conventional data base systems. Further, the TIGER file's spatial characteristics cannot be handled by data base management systems which do not understand the concept of distance. Therefore, the TIGER file structure is designed as a series of interlocking files. These are called lists and directories. The lists are characterized by random storage of their elements while the directories have their elements stored in a particular order. Each element can be considered as a record in a list or directory. The elements contain data items and/or index pointers to other elements or to elements in other files. These pointers give the TIGER file structure its interlocking nature and provide a means for rapid access and management of the total file, both spatially and in more conventional fashions.

The implicit relationships built in to the relational data-base structure of the TIGER Files are established by record types, defining separate tables linked together by a common data field. Within each record type are explicit definitions of attributes and geometric characteristics necessary to produce maps and to define topological relationships.

TIGER/Line Files

The first TIGER product provided by the Census Bureau for use by local government was the TIGER/Line File, which became available in 1989. Each TIGER/Line File (one for each county) contains geographic features (roads, rivers, railroads, etc.) that were scanned from the 1:100,000 scale USGS maps for areas not covered by the 1980 DIME Files. (Geographic features in areas covered by the 1980 DIME Files were "fit" into the TIGER map from the DIME File information.) The TIGER/Line Files also contain census geographic area codes (block numbers, census tracts, etc.), which were obtained from the Census Bureau's Boundary and Annexation Survey (BAS); address ranges for those areas covered by the 1980 DIME Files; and latitude/longitude coordinates for all points (Bureau of the Census, 1989).

There are six different record types in the relational structure of the TIGER/Line File, forming six separate tables that are related to each other by a unique record number for each line segment. The first two record types contain the line segment identification and cartographic information that are used to produce maps (see Figure 5.13). The remaining four record types contain the geographic information used for geocoding and geoprocessing applications. The following descriptions of these tables are summarized from the "TIGER/Line Prototype Files, 1990: Technical Documentation" (Bureau of the Census, 1989).

BASIC DATA RECORD (RECORD TYPE 1)

This table contains a record for each unique line segment on the map and the Census Bureau geography related to it. Latitude and longitude coordinates are provided for each segment endpoint, as are address ranges (for areas covered by the 1980 DIME Files) and geographic area codes for each side of the segment. (Table 5.7 lists all of the data elements contained in the basic data record.)

SHAPE COORDINATE PAIRS (RECORD TYPE 2)

This table contains up to ten additional points (expressed in latitude and longitude coordinates) for line segments that are not straight lines. It allows the display of a curved line segment, for example, without creating separate line segments (Type 1 records) for each arc of the segment. Type 2 records are related to Type 1 records by the record number of the Type 1 record. Type 1 records that are straight lines do not have corresponding Type 2 records. (Table 5.8 lists all of the data elements contained in the Shape Coordinate Pairs File.) For example, Line Segment L2 in Figure 5.14 represents a curved street segment running from a street intersection at P1 to another street intersection at P2. The TIGER/Line File for Figure 5.14 contains one basic data record (Record Type 1) for Line Segment L2 and latitude and longitude coordi-

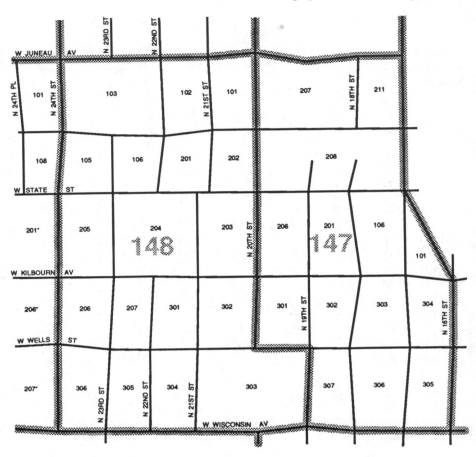

FIGURE 5.13. A portion of a plot from the TIGER/Line File showing two census tracts (147 and 148) and their block numbers.

nates for its endpoints, P1 (FROMLAT and FROMLONG) and P2 (TO-LAT and TOLONG).

Since Line Segment L2 is not a straight line from P1 to P2, its basic data record (Record Type 1) is related by Record Number to a shape coordinate pairs (Record Type 2) record which contains the additional latitude and longitude coordinate points (for P4–P13) to define the curvature of the line segment. This eliminates the need for 11 separate line segments to be defined in the Basic Data Record (as would be the case with the DIME File).

1980 CENSUS GEOGRAPHIC AREA CODES (RECORD TYPE 3)

This table contains records for each block identified in the 1980 DIME Files and associates higher-level geography (census tract, place code,

Table 5.7. Data elements in the TIGER/Line File basic data record (Record Type 1).

Name	Data type	Size	Character position Beg	Character position End	Description
RECTYP	N	1	1	1	Record type (Value = "1")
FILECDE	N	5	2	6	File code
RECNUM	N	8	7	14	Record number
SIDECDE	N	1	15	15	Single side street code (used to signify that data exists only for one side of the feature segment. Normally, this is a county boundary)
SOURCE	A	1	16	16	Source code
ID	A	38	17	54	Feature identification
DIRPRE	A	2	17	18	Feature direction, prefix
FEANME	A	30	19	48	Feature name
FEATYP	A	4	49	52	Feature type
DIRSUF	A	2	53	54	Feature direction, suffix
CFCC	A	3	55	57	Census feature class code
FRADDL	A	11	58	68	Left from address
TOADDL	A	11	69	79	Left to address
FRADDR	A	11	80	90	Right from address
TOADDR	A	11	91	101	Right to address
FRIADDFL	N	1	102	102	Left from imputed address flag
TOIADDFL	N	1	103	103	Left to imputed address flag
FRIADDFR	N	1	104	104	Right from imputed address flag
TOIADDFR	N	1	105	105	Right to imputed address flag
ZIPL	N	5	106	110	ZIP Code (left)
ZIPR	N	5	111	115	ZIP Code (right)
AIRL	N	4	116	119	American Indian Reservation Alaska Native Village Statistical Area Tribal Designated Statistical Area Tribal Jurisdiction Statistical Area Code left
AIRR	N	4	120	123	American Indian Reservation Alaska Native Village Statistical Area Tribal Designated Statistical Area Tribal Jurisdiction Statistical Area Code right
ANCL	N	2	124	125	Alaska Native Regional Corporation Code (left)
ANCR	N	2	126	127	Alaska Native Regional Corporation Code (right)
ST90L	N	2	128	129	FIPS State Code (left)
ST90R	N	2	130	131	FIPS State Code (right)
CO90L	N	3	132	134	FIPS County Code (left)
CO90R	N	3	135	137	FIPS County Code (right)

Table 5.7. Data elements in the TIGER/Line File basic data record
(continued)

Name	Data type	Size	Character position Beg	End	Description
MCD90L	N	3	138	140	Minor Civil Division (MCD)/ Census County Division (CCD) Code (left)
MCD90R	N	3	141	143	Minor Civil Division (MCD)/ Census County Division (CCD) Code (right)
PLCDE90L	N	4	144	147	Place Code (left)
PLCDE90R	N	4	148	151	Place Code (right)
CTBNA90L	N	6	152	157	Census Tract/BNA Code (left)
	N	4	152	155	Basic number
	N	2	156	157	Suffix
CTBNA90R	N	6	158	163	Census tract/BNA Code (right)
	N	4	158	161	Basic number
	N	2	162	163	Suffix
BLK90L	A	4	164	167	Tabulation block number (left)
	N	3	164	166	Collection block number
	A	1	167	167	Tabulation suffix
BLK90R	A	4	168	171	Tabulation block number (right)
	N	3	168	170	Collection block number
	A	1	171	171	Tabulation suffix
FROMLONG	N	9	172	180	From longitude (implied 6 decimal places)
FROMLAT	N	8	181	188	From latitude (implied 6 decimal places)
TOLONG	N	9	189	197	To longitude (implied 6 decimal places)
TOLAT	N	8	198	205	From latitude (implied 6 decimal places)

etc.) to each block. Linked to the line segments of the Basic Data Record by record number, these geographic areas represent 1980 geography only. New blocks and other geographic areas created since 1980 are identified within the Basic Data Record (the Block Tabulation Suffix field). (Table 5.9 lists all of the elements in the 1980 Census Geographic Area Codes table.)

INDEX TO ALTERNATE FEATURE NAMES (RECORD TYPE 4)

Any line segment in a Basic Data Record that is known by more than one name has a Record Type 4, which identifies a code (Alternate Feature Code) representing the alternate name of the segment's Feature Name. This code relates the record to the Feature Name List (Record Type 5), where the name is stored. For example, a line segment representing "Main Street" may have an Alternate Feature Name if it is

Name	Data type	Size	Character Beg	End	Description
RECTYP	N	1	1	1	Record type (value = "2")
FILECDE	N	5	2	6	File code
RECNUM	N	8	7	14	Record number
RTSQ	N	3	15	17	Record sequence
LONG1	N	9	18	26	Longitude coordinate for first shape point
LAT1	N	8	27	34	Latitude coordinate for first shape point
.	.	.			.
.	.	.			.
.	.	.			.
LONG10	N	9	171	179	Longitude coordinate for tenth shape point
LAT10	N	8	180	187	Latitude coordinate for tenth shape point

FIGURE 5.14. A curved street segment (L2) that is defined in the TIGER/Line File as one basic data record (Record Type 1) related to one shape coordinate pairs record (Record Type 2), which contains latitude and longitude coordinates for each of the 10 intermediate points (P4 through P13) defining the curvature of the street segment.

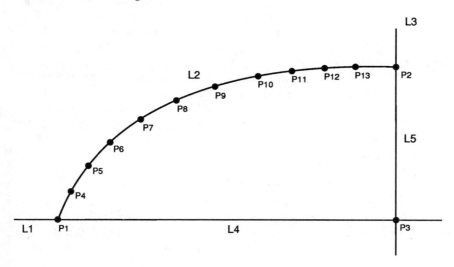

175

Table 5.9. Data elements in the TIGER/Line File 1980 Census geographic area codes (Record Type 3).

Name	Data type	Size	Character position Beg	End	Description
RECTYP	N	1	1	1	Record type (value = "3")
FILECDE	N	5	2	6	File code
RECNUM	N	8	7	14	Record number
ST80L	N	2	15	16	1980 FIPS State Code (left)
ST80R	N	2	17	18	1980 FIPS State Code (right)
CO80L	N	3	19	21	1980 FIPS County Code (left)
CO80R	N	3	22	24	1980 FIPS County Code (right)
MCD80L	N	3	25	27	1980 Minor Civil Division/Census County Division Code (left)
MCD80R	N	3	28	30	1980 Minor Civil Division/Census County Division Code (right)
PLCDE80L	N	4	31	34	1980 Place Code (left)
PLCDE80R	N	4	35	38	1980 Place Code (right)
CTBNA80L	N	6	39	44	1980 Census Tract (left)
	N	4	39	42	Basic number
	N	2	43	44	Suffix
CTBNA80R	N	6	45	50	1980 Census Tract (right)
	N	4	45	48	Basic number
	N	2	49	50	Suffix
BLK80L	N	3	51	53	1980 block (left)
BLK80R	N	3	54	56	1980 block (right)

also known by a State or U.S. Highway designation. This Alternate Feature Name record is linked to its Basic Data Record by its Record Number and to the Feature Name List by its Alternate Feature Code. (Table 5.10 lists the elements contained in the Index to Alternate Feature Names.) Main Street in Figure 5.15, for example, is also designated as U.S. Highway 41. There is one Basic Data Record (Record Type 1) for each street segment (L1–L4) running from 1st Street to 5th Street. Each Basic Data Record contains a Feature Name of "Main Street."

Since Main Street has an alternate name (U.S. Highway 41), the Index to Alternate Feature Names (Record Type 4) will also have a record for each street segment (L1–L4). To minimize file storage space, these records have a code (Alternate Feature Code) for the feature rather than the name itself.

FEATURE NAME LIST (RECORD TYPE 5)

All unique Feature Names (street names, river names, railroad names, etc.) and their Alternate Feature Names are contained in this file. Each Feature Name is given an Alternate Feature Code that links it to an Alternate Feature Name record and is used as shorthand for the alter-

Table 5.10. Data elements in the TIGER/Line File index to alternate feature names (Record Type 4)

Name	Data type	Size	Character position Beg	End	Description
RECTYP	N	1	1	1	Record type (value = "4")
FILECDE	N	5	2	6	File code
RECNUM	N	8	7	14	Record number
RTSQ	N	3	15	17	Record number sequence
FEAT1	N	8	18	25	Alternate Feature 1 Code
FEAT2	N	8	26	33	Alternate Feature 2 Code
FEAT3	N	8	34	41	Alternate Feature 3 Code
FEAT4	N	8	42	49	Alternate Feature 4 Code
FEAT5	N	8	50	57	Alternate Feature 5 Code

nate names. (Table 5.11 lists the elements that are contained in the Feature Names List.)

ADDITIONAL ADDRESS RANGES AND ZIP CODE DATA (RECORD TYPE 6)

When the address range along a street segment in the Basic Data Record cannot be presented as a single address range, Type 6 records are used to identify the additional address ranges that also are related to the segment. Sometimes, for example, a specific building will have a street address number that does not fit within the low and high numbers of the address range, yet will still be located on that street segment. The 711 Building in Figure 5.16 has an address of 711 Main Street but is located on the street segment having the address range of 600–699 Main Street.

Since address anomalies such as shown in Figure 5.16 do occur in local government (usually through a legal process such as an ordinance), the TIGER/Line File Type 6 record was devised to eliminate the

FIGURE 5.15. Four street segments (L1 through L4) having an Alternate Feature Name (U.S. Highway 41).

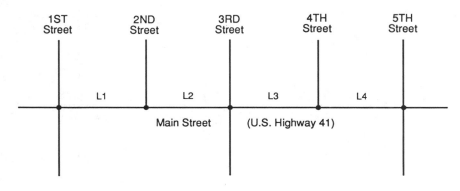

Table 5.11. Data elements in a TIGER/Line File feature name list (Record Type 5).

Name	Data type	Size	Character position Beg	End	Description
RECTYP	N	1	1	1	Record type (value = "5")
FILECDE	N	5	2	6	File code
FEAT	N	8	7	14	Alternate Feature Code
ID	A	38	15	52	Feature identification
DIRPRE	A	2	15	16	Feature direction, prefix
FEANME	A	30	17	46	Feature name
FEATYP	A	4	47	50	Feature type
DIRSUF	A	2	51	52	Feature direction, suffix

need to subdivide a segment into smaller segments (and, therefore, Basic Data Records). The Basic Data Record (Record Type 1) for the 600–699 Main Street segment is related to a record in the Additional Address Range and ZIP Code Data (Record Type 6) Table, which contains an address range of 711–711 Main Street. Thus, the Type 1 record contains the address range 600–699 Main Street, and the Type 6 record contains the address range 711–711 Main Street for the same street segment (the two records are related by having the same record number).

Type 6 records are also used to identify the address ranges within street segments where ZIP Code boundaries cut through the segments. Instead of creating two segments to accommodate a ZIP Code boundary that cuts through a street segment, there is only one street segment (Basic Data Record), but it is linked to additional records in the Additional Address Range and ZIP Code Data Table. Figure 5.17, for ex-

FIGURE 5.16. A building with an address not within the normal address range of a street segment. The TIGER/Line File provides a record in the Additional Address Range & ZIP Code Data table (Record Type 6) to accommodate this situation.

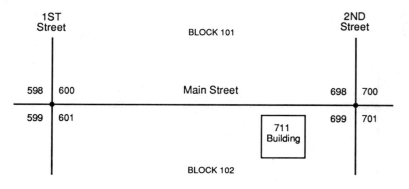

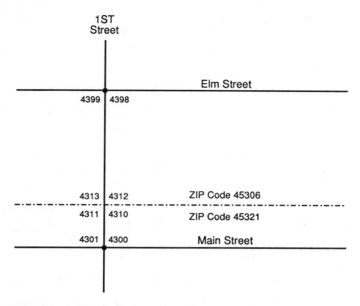

FIGURE 5.17. A ZIP Code boundary that cuts through a street segment.

ample, shows a street segment along 1st Street that is divided into two ZIP Code areas (45306 and 45321).

Since ZIP Code 45321 in Figure 5.17 includes all properties on the north side of Main Street, some properties on 1st Street between Main Street and Elm Street are in Zip Code 45321, and others are in Zip Code 45306. Since there is only one street segment on 1st Street between Main and Elm (and, therefore, only one Record Type 1), there are two Additional Address Range and ZIP Code Data (Record Type 6) records for the segment. These additional records (one for each ZIP Code) have the same record number as the Type 1 record for the segment.

Table 5.12 lists the elements that are contained in the Additional Address Ranges and ZIP Code Data Table.

For local governments covered by the 1980 DIME Files, there is not a significant difference between the 1980 DIME File and the 1990 TIGER/Line File other than the addition of new geographic features created during the 10 years between decennial censuses—information that local governments have already recorded as part of their ongoing public works construction activities. While Record Types 4, 5, and 6 do help correct problems with the DIME Files in representing certain anomalies in local geography, the investment in computer programs and data-base structuring to use the TIGER/Line File will be certain to weigh heavily in the decision to switch from a DIME-based to a TIGER-based geoprocessing system (the Census Bureau provides no programs

Table 5.12. Data elements in the TIGER/Line File additional address range and ZIP Code data (Record Type 6).

Name	Data type	Size	Character position Beg	End	Description
RECTYP	N	1	1	1	Record type (value = "6")
FILECDE	N	5	2	6	File code
RECNUM	N	8	7	14	Record number
RTSQ	N	3	15	17	Record sequence number
FRADDL	A	11	18	28	From address (left)
TOADDL	A	11	29	39	To address (left)
FRADDR	A	11	40	50	From address (right)
TOADDR	A	11	51	61	To address (right)
FRIADDFL	N	1	62	62	From imputed address flag (left)
TOIADDFL	N	1	63	63	To imputed address flag (left)
FRIADDFR	N	1	64	64	From imputed address flag (right)
TOIADDFR	N	1	65	65	To imputed address flag (right)
ZIPL	N	5	66	70	ZIP Code (left)
ZIPR	N	5	71	75	ZIP Code (right)

for local governments to use in processing the TIGER Files). Record Types 1 and 2 will allow an urban geographic information system to create maps of line segments; however, most local records are much more accurate and up to date than the 1:100,000 scale maps used by the USGS to create the digital map records. The Census Bureau itself recommends that these maps can be used for statistical purposes but are not suitable for the high-precision engineering and property transfer applications required by local governments.

The primary beneficiaries of the TIGER Files are the U.S. Census Bureau, which no longer has inconsistencies between its enumeration maps and its geographic base file, and organizations that require digital cartographic information covering geographic areas as large as states or the entire country. Local governments that did not use the DIME File for geoprocessing or do not now use a geographic information system for processing digital cartographic records will find that the TIGER Files can be useful for many geographic information system applications.

GIS County Exercise 2: Geographic Base Files

GIS County needs a Geographic Base File to use in its geographic information system for processing geographic information. The land in GIS County is being developed and needs an addressing scheme so that the parcels of land can be located. Building inspectors, tax assessors, water meter readers, public health nurses, local elected officials, delivery persons—all need to know where the properties are and what information about them is available.

The county's GBF will help collect the information and integrate it with other information that is available in computerized records. Your job is to build a GBF that can be computerized. Figure E5.1 identifies all the address ranges for the streets in the county. Figure E5.2 is from the Census Bureau and identifies the block numbers for the county.

Using these maps and those from the previous exercise in Chapter · 4, you can build a GBF to prepare for the computerization of all the property records in GIS County. You must record all the possible addresses in the county and identify in which block and district they are. Remembering that right and left polygons are defined as if you were at the low end of the address range facing the high end, use the following format:

Segment House No. House No.   Dir Street
  Label   Low   High
  name   Right Left   Right   Left
          Block Block District District

Add the new points and areas to the point table and area table and then solve the following five problems using only the GBF you have built. After you solve each problem and before you work on the next one, write in your own words the procedure you used to solve it.

1. Identify all the boundaries of GIS County. List the streets and then the address ranges of all the blocks to create the following table:

	House no.	House no.	
Street name	Low	High	Segment label

2. List all the block numbers in each district:

District	Blocks
A	
B	
C	
D	
E	

3. Identify all the boundaries of each district. List the streets and address ranges of all the blocks to create the following table:

		House no.	House no.	
District	Street name	Low	High	Segment

4. List all the address ranges for properties in District D:

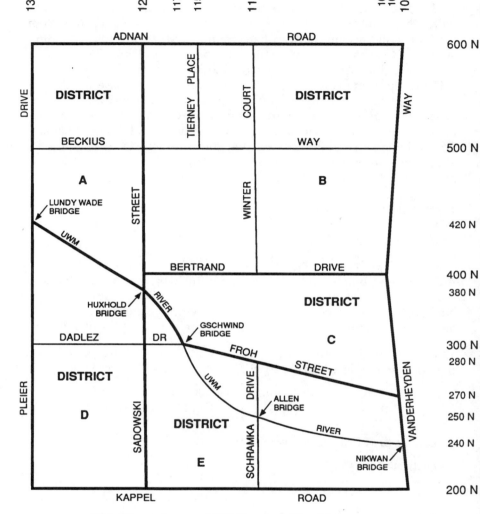

E5.1. General map of GIS County with address ranges.

House no. Low	House no. High	Direction	Street name

5. Identify the blocks of the following addresses (note: odd house numbers are in the left block and even house numbers are in the right block):

Address   Block
240 N. Pleier Dr.
1170 W. Dadlez Dr.
412 N. Winter Ct.
575 N. Tierney Pl.
1060 W. Beckius Wy.
1277 W. Adnan Rd.
1051 W. Bertrand Dr.

E5.2. GIS County block identifiers and block centroid identifiers.

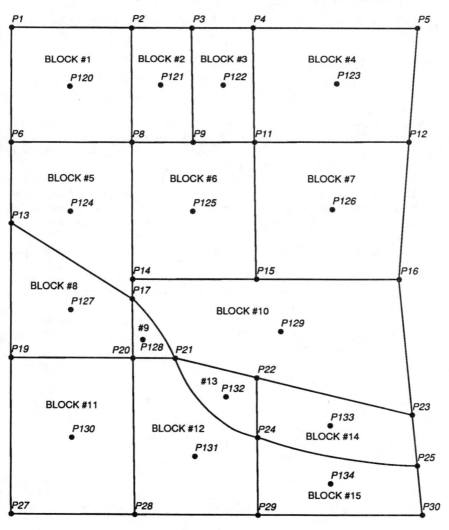

References

Bureau of the Census (1980), "Geographic Base File/DIME (GBF/DIME), 1980: Technical Documentation," prepared by Data Access and Use Staff, Data User Services Division, Washington: The Bureau.

Bureau of the Census (1989), "TIGER/Line Prototype Files, 1990: Technical Documentation," prepared by the Data Access and Use Staff, Data User Services Division, Washington: The Bureau.

Broome, Frederick R. (1984), "TIGER Preliminary Design and Structure Overview: The Core of the Geographic Support System for 1990," presentation at the 1984 Annual Meeting of the American Association of Geographers, Washington, D.C.

Fletcher, David R. (1989), "Integrating Network Data into a Transportation Oriented Geographic Information System," Third Annual Midwest/Great Lakes ARC/INFO User Conference, October 17–19, 1989, Madison, Wisc.

ADDITIONAL READINGS

Boudriault, G. (1987), "Topology in the TIGER File," *Proceedings: Eighth International Symposium on Computer Assisted Cartography*, (pp. 258–63).

Bureau of the Census (1985), "TIGER Tales," Washington, D.C.

Fletcher, David (1987), "Modelling GIS Transportation Networks," URISA 24th Annual Conference Proceedings, Fort Lauderdale, Fla.

Kinnear, C. (1987), "The TIGER Structure," *Proceedings: Eighth International Symposium on Computer Assisted Cartography*, (pp. 249–57).

McKenzie, B.Y., and LaMacchia, R.A. (1987), "The U.S. Geological Survey— U.S. Bureau of the Census Cooperative Digital Mapping Project: A Unique Success Story," unpublished paper presented at the Fall (1987) Meeting of the American Congress on Surveying and Mapping, Reno, Nevada.

Marx, R. W. (1986), "The TIGER System: Automating the Geographic Structure of the United States Census," *Government Publications Review*, Vol. 13, (pp. 181–201).

Marx, R.W. (1988), "The TIGER System: Six Years to Success," *Proceedings of the 13th International Cartographic Conference*, Morelia, Michoacan, Mexico, 12–21 October, 1987, Vol IV, (pp. 633–45).

Trainor, T.F. (1986), "Attribute Coding Scheme: Identification of Features and the U.S. Census Bureau's TIGER System," *Proceedings: AUTO CARTO London*, Vol. 1, (pp. 117–26).

6

Land Records Information

There is a critical need for a better land-information system in the United States to improve land-conveyance procedures, furnish a basis for equitable taxation, and provide much-needed information for resource management and environmental planning.*

Chapter 2 described land records as the basis upon which geographic information is related to physical locations on the earth. In a geographic information system, this is accomplished by the use of three types of referencing systems:

1. *Cartographic references*—Maps that define spatial relationships among cartographic features;
2. *Coordinate systems*—Coordinates that define the positions of features on the surface of the earth so that they can be accurately represented on maps; and
3. *Location identifiers*—Codes that are used to link attributes of physical entities to their cartographic features.

While local governments have a variety of choices to make among these three types of reference systems (such as the two examples shown in Figures 2.8 and 2.9 of Chapter 2), at least one choice from each of the three types must be present in a GIS in order for location-related attribute data to have spatial meaning to local government officials. This is because *maps* are needed to identify where one feature is in relation to another feature; a *continuous coordinate system* is needed to relate separate maps to each other; and *location identifiers* are needed to relate data about physical entities to the objects that represent them on maps.

---

*Reprinted from "Need for a Multipurpose Cadastre," 1980, with permission from the National Academy Press, Washington DC., p. 1.

Cartographic References

As explained in Chapter 4, maps contain points, lines, polygons, symbols, and text in a spatial context that allows a visual understanding of how these cartographic objects are related to each other. When two lines intersect on a map, for example, we know that the features they represent are related to each other at the point of intersection. Similarly, when a street name is printed above a line representing a street, we know the name of that street.

In an urban geographic information system, a *base map* contains the cartographic information that is common among all the different users of maps. The base map consists of either *cadastral* information (the legal identification of features) or *planimetric* information (the physical identification of features), or it may consist of a combination of both cadastral and planimetric information. The difference between cadastral and planimetric information is significant, and the decision a local government makes concerning which to choose to form its base map is a major decision that can affect both the use of the system and the cost of the task to create the base map. This is because cadastral maps are based upon the legal definition of parcels of land and planimetric maps are based on physical features that can be seen on the land. For reasons that will be explained in this chapter, the two types of maps do not usually agree when locating features that are common to both.

Cadastral Maps

In local government, cadastral records form the basis upon which property is conveyed, assessed for taxation, and managed for delivering public services. Cadastral records consist of maps, documents, manual and computerized files, and other legal and official instruments that contain information about the cadastre—legal information about property. A definition of the cadastre has been presented by the Panel on a Multipurpose Cadastre of the Committee on Geodesy in the National Research Council's Assembly of Mathematical and Physical Sciences (Panel on a Multipurpose Cadastre). The Panel's definition reads:

> A cadastre may be defined as a record of interests in land, encompassing both the nature and extent of those interests.*

An "interest in land" is more than just legal ownership because many conditions can exist that allow others besides the owners to have some legal claim to properties that can restrict either its use or its conveyance. A financial institution, for example, has a legal claim to a prop-

*Reprinted from "Need for a Multipurpose Cadastre," 1980, with permission from the National Academy Press, Washington DC., p. 5.

erty owned by someone who has obtained a mortgage from the institution. Before the property is sold, the mortgage must be paid off, and therefore the financial institution has an interest in the property. A utility company has an interest in a property when there is a utility easement on the property that restricts its use. Other people and organizations that have construction liens, loans secured by property, pending court actions, special assessments, and other legal restrictions such as zoning laws, claims of legal right-of-way, etc. all have interests in land owned by someone else.

The United States has an elaborate system to help ensure that legal claims of interest in property are either transferred or satisfied when ownership of the property changes. This system is generally referred to as a *land-title recordation system* and is based upon the official recording of such documents by local government (usually a County Recorder or Register of Deeds or a City or Town Clerk with similar responsibilities). These documents are official or legal instruments claiming interests in property and are considered "recorded" when the local government receives them, stores them, and records them in "registers."

Conveying property requires not only the legal description of the parcel (where it is and what its shape is), but also a methodology for identifying all the legal claims of interest in the parcel. The records local governments maintain for administrating the land-title recordation function include:

- Copies of deeds that contain legal descriptions of properties and the grantor/grantee names;
- Plat maps, which are a graphical representation of the legal descriptions of property (see Figure 3.1);
- Copies of the legal documents claiming interests in properties;
- An index to these documents, usually ordered by the names of grantors and grantees;
- An index to the maps, usually by block identification.

The primary purpose of assessing property is to provide for a fair and equitable taxation of the land for producing the revenue needed to run local government. Fair and equitable taxation requires that properties having similar characteristics (size, location, type of building, age, income-producing potential, etc.) also have similar taxes levied against them. As in land-title recordation, property assessments are the responsibility of local government in the United States. Depending upon the state and the region within the state, property assessments are performed either by the county, the township, or the municipality. These units of government perform the following functions in assessing property:

- Locating and describing properties;
- Appraising or estimating the values of properties;
- Keeping records that link owners to their properties;
- Designating the official value of properties for taxation.

The records most tax assessors maintain and use in performing these functions include tax maps (such as shown in Figure 3.1), property record cards containing detailed notes, characteristics, and sketches describing each property, and computerized property characteristic data bases.

The tax maps used by local governments to assist them in their land-title recordation and property assessment responsibilities are "cadastral maps," which are created and maintained through the legal authority and responsibility of local government. When a local government implements a geographic information system, it may choose to use these maps (or the legal descriptions from which the maps were created) to build its cadastral base map. This can be accomplished by digitizing or scanning the maps, or it can be accomplished by using computer programs to perform coordinate geometry (COGO) functions, which create digital records directly from the legal descriptions of the properties. Whichever method is used, this digital conversion process is time-consuming and frustrating because the maps and their legal descriptions are not error-free (survey methods and drafting standards—and the training and ability of the people who use them—have varied considerably over the 200 or so years that property has been recorded in the United States). Chapter 7 discusses in more detail the factors that affect the base map conversion process.

Planimetric Maps

Planimetric maps are maps created from field surveys or aerial photographs. Since property lines, easements, and rights-of-way are defined by legal descriptions, they cannot be identified by sight and, consequently, do not appear in planimetric maps. Physical features—entities that can be seen on the ground—are recorded on planimetric maps: Curb lines, roadways, sidewalks, street intersections, rivers, lakes, trees, manhole covers, fire hydrants, buildings, bridges, even fence lines (which may or may not be located on property lines) are some of the entities that can be located on planimetric maps because they can be identified during field surveys or on photographs taken from airplanes.

Some local governments use aerial photographs to build the base map for their geographic information system, thus producing a digital cartographic data base of planimetric information. Such a planimetric data base contains information that is similar to that contained in a cadastral data base (streets, intersections, rivers, etc.); however, it is missing the information necessary to map parcel boundaries. On the other hand, planimetric data bases contain other information not recorded in cadastral data bases such as: curb lines (defining pavement width), building locations, and utility facilities on the ground (manholes, fire hydrants, transformers, street lights, and power, telephone, and Cable TV poles). Through the stereo digitizing process described in Chapter 2, ground elevations can also be obtained from aerial photographs to produce *topographic maps* with elevation contour lines. This information may be desirable in jurisdictions that use a geographic information system to produce "cut and fill" diagrams for roadway construction and site preparation for industrial parks or other major public construction projects.

Figure 6.1 shows a planimetric map of one city block. It contains physical features that were identified from an aerial photograph: curb lines, sidewalks, building outlines, fences, and street lights. It also contains spot elevation points and elevation isobars forming contour lines of the topography of the land.

An advantage planimetric maps created from aerial photographs have

FIGURE 6.1. A planimetric map of one block created from an aerial photograph.

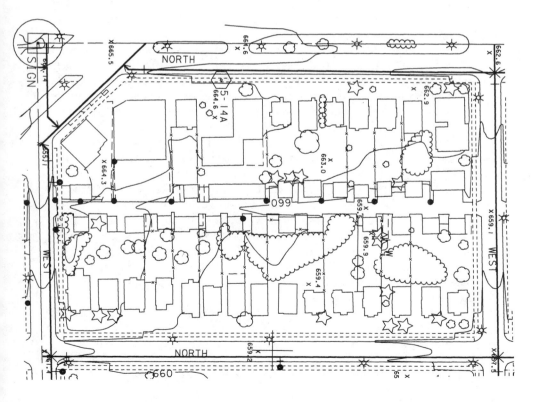

over cadastral maps for creating a digital data base in a geographic information system is that the aerial photographs fit together in a mosaic pattern, thus producing a continuous map of the entire jurisdiction. Joining map sheets of cadastral information into a continuous map of the entire jurisdiction creates gaps and overlaps where the map sheets meet because of the variations in surveying techniques and conditions over the hundreds of years during which the maps were created and updated. Since aerial photographs are taken at a specific point in time, no such gaps or overlaps appear in the planimetric maps—even when they are subdivided into separate map sheets.

There are two conditions, however, that cause distortions in the relative locations of features recorded in aerial photographs that must be corrected before using the photos as sources for cartographic data in a digital data base. First, the surface of the earth is uneven, and, when it is projected onto the flat surface of a photograph, distortions in the relative distances between features occur because they are at different elevations (and, consequently, at different scales). Figure 6.2 shows this distortion between the flat surface of a photograph (containing the points labeled a–e) and the uneven surface of the earth (containing physical entities at points labeled A–E). The actual distance on the ground (known as "ground truth") between the two entities at points D and E, for example, is not accurately represented on the photograph by the points d and e because the scale along de is larger than the scale along ab (since D and E have higher elevations than A and B).

The second condition that causes distortion in aerial photographs is the tilt of the airplane (and, hence, the camera) at the time the photograph is taken. Because the airplane cannot maintain an absolutely perpendicular angle between the camera and the ground when the shutter is released, the resulting photograph is not parallel to the ground and, thus, contains distortions in distances between features. Figure 6.3 shows the distortion on a photograph caused by the tilt of the camera in an airplane over level terrain. The tilted surface of the photograph (represented by the line containing points a–e) inaccurately records the locations of entities on the surface of the earth (represented by points A–E). If the photograph were parallel to the surface of the earth, the locations of the two points A and B would be recorded on the level surface at points a' and b'. Instead, the tilted surface of the photograph records their locations at points a and b. The photograph contains distortions because the distance between a and b on the photograph is larger than the distance between a' and b', which accurately represents the distance between A and B.

The process of correcting the distortion between the actual distance on the earth and the measured distance on a photograph of the earth is called *orthogonal rectification*. Orthogonal rectification is a mathematical manipulation of the locations of features identified on a photograph using the elevation of the airplane when the photograph was taken

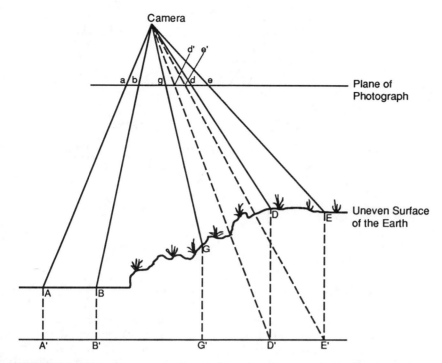

FIGURE 6.2. Distortions on the flat surface of a photograph (the horizontal plane containing points *a* through *e*) caused by the uneven surface of the earth (the points labeled *A* through *E*). Because the surface of the earth at distance *DE* is at a higher elevation than distance *AB*, their scales on the photograph are different. Thus, the scale of the length *de* on the photograph is different than the scale of the length *ab*. The scale of the length *d'e'* is at the same scale as the length *ab*, but *d'* and *e'* are not recorded on the photograph.

Illustrations from *Surveying Theory and Practice (Sixth Edition)*, by R. E. Davis, F. S. Foote, J. M. Anderson, and E. M. Mikhail (1981), McGraw-Hill, New York, NY.

and the actual locations and elevations of certain known features on the earth. When these known features are located by surveyors and the distances between them are measured on the ground, then the actual measurements can be used to adjust all of the measurements on the photographs. Photographs rectified in this manner are called *orthophotographs* (or ortho photos). The maps created from the ortho photos contain measurements that represent actual ground truth and thus are used to create planimetric base maps in some urban geographic information systems.

Planimetric Base Maps with Cadastral Overlays

Some jurisdictions use a combination of planimetric information and cadastral information to create their base map for a geographic infor-

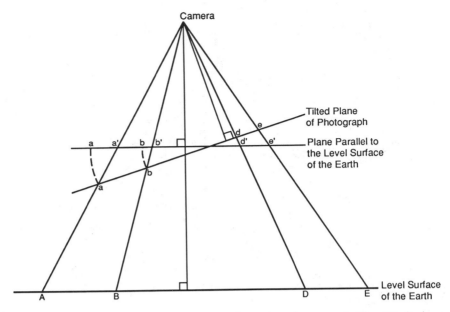

FIGURE 6.3. Distortions on the flat surface of a photograph (the tilted plane containing points *a* through *e*) caused by the tilt of the camera in the airplane over level terrain (represented by the points *A* through *E*). The distance between points *A* and *B* is recorded on the photograph as the distance between points *a* and *b*, when, in fact, the accurate distance would be that between points *a'* and *b'*. Illustrations from *Surveying Theory and Practice (Sixth Edition)*, by R. E. Davis, F. S. Foote, J. M. Anderson, and E. M. Mikhail (1981), McGraw-Hill, New York, NY.

mation system. This is common among jurisdictions that do not maintain cadastral maps of land parcels (plat maps) or that have such maps, but they are either inaccurate, out of date, or physically unusable for digitizing or scanning. In these jurisdictions, planimetric maps are created from orthophotographs, and the parcel boundaries are added from the legal descriptions of the parcels, using reference points common to both sources. These references are usually physical features identified on the orthophotograph and also referenced in a parcel's legal description: the centerline of a street, a railroad, the shoreline of a river or lake, and other physical entities that, while too small to see on an aerial photograph, have been temporarily marked with a larger object (known as a *target*). Locating parcels in relation to these common reference points is often difficult and, in a densely populated area, very time-consuming because of the large number of parcels and the small number of reference points common to both map sources.

The State of North Carolina formed the North Carolina Land Records Management Program (LRMP) in 1977 to assist all 100 counties in the state create base maps by combining planimetric maps with cadastral records from either tax maps or the legal descriptions of parcels from their deeds. The process is described by Donald P. Holloway, Director of the North Carolina Land Records Management Program, in a paper he presented at the 1983 annual conference of the Urban and Regional Information Systems Association. After creating a base map of planimetric information from orthophotographs, the cadastral information was added as described below:

> The compilation and production of a cadastral map is as challenging as it is essential. The initial step is done by researchers who comb the Register of Deeds' books for every deed pertaining to any part of the land covered by the subject map. Once this "deed research" is completed, the next phase is the input of all boundary information into an automated plotter, which computes and draws an exact, to-scale figure of the parcel. These figures are transferred by tracing to a "work copy" of the respective base map, keying the parcel corners and lines to terrain features wherever possible. Typically, these "parcel figures" do not fit together precisely on the map draft, and so the third step is that of reconciliation of any gaps or overlaps between adjoining parcel boundaries. This reconciliation may be done by references to "use lines" or by "field checking." "Use lines" such as fences, roads and streams are often relied on as a visible means of determining where adjacent property owners intend their common boundary to be. Mappers will go "to the field" to consult with neighboring property owners to inform them of major discrepancies as to boundaries in their deeds and to attempt to arrive at the most accurate boundaries possible under the circumstances. (p. 186)

Forsyth County (NC), one of the first counties to participate in this program, found this process to be very difficult in spite of having the cadastral information for all 140,000 parcels already mapped prior to combining them with the planimetric base map. Eunice H. Ayers, then Register of Deeds of Forsyth County, describes this difficulty in a paper published by the University of Wisconsin—Madison Institute for Environmental Studies (1984):

> The most difficult single phase in the project has been to complete the production of the graphic base (map)—identifying and resolving discrepancies between the graphic parcel descriptions digitized from tax maps and orthophotographic base maps. This process of rectification has been long, difficult, and frustrating.   p. 298

She goes on to observe that:

> In compiling the land records, it is clear for the first time what mistakes and errors have existed in those records for years.   p. 298

A graphic representation of the problems encountered by overlaying cadastral information onto a planimetric map is given in Figure 6.4. This city block is the same as that shown in Figure 6.1, with parcel boundaries added to the planimetric base map. These boundaries were

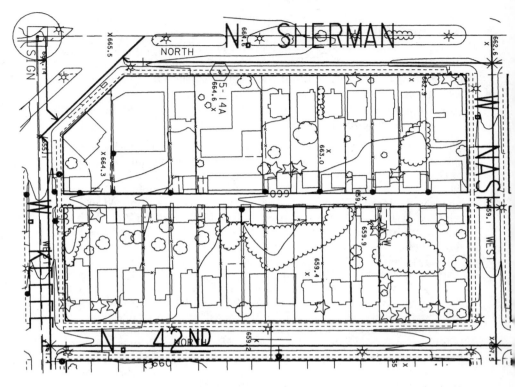

FIGURE 6.4. Cadastral map features that do not precisely match the physical features represented on a planimetric map. Here, property lines do not match fence lines and cut a portion of building outlines when a cadastral map is overlaid onto the planimetric map of Figure 6.1.

digitized from tax plat maps created from the legal descriptions of the parcels. Notice that some parcel boundaries (shown as solid lines) are overlaid onto a portion of the buildings on the parcels and that fence lines do not match up with parcel boundaries.

Base Map Accuracy

Whether cadastral maps or planimetric maps (or a combination of both) are used to create the base map of an urban geographic information system, they must be free of "mistakes and errors" in order to support the many different applications of GIS technology at the local government level. Depending on the applications in which an urban GIS is used, the accuracy of map information can influence policy decisions (such as those based upon statistics related to area—liquor license density, for one); legal decisions (such as land ownership or zoning regulations); or even life-threatening situations (such as inadvertently cutting a buried power line or gas main).

The only way to determine if a map is accurate is to check to see if the features on the map are positioned in the same manner as their true locations on the earth. While a discussion of *vertical accuracy* of maps (how accurate elevations are represented on topographic maps) is beyond the scope of this book, *horizontal accuracy* deserves considerable attention since the accuracy of locations on a map has a significant impact not only on how a GIS is used, but also on the cost and effort required to create its base map.

A map is horizontally accurate when a feature on the map has the same relative location as the entity it represents on the earth. A map is accurate, for example, when it can be used to find an underground water main without digging up half the street. It is even more accurate when the amount of digging required to find the main is minimized. In order to find a feature in the real world by using a map, the location of the feature must be related to some reference point that can be found and from which measurements can be made (before digging for the water main, measurements are made from a physical entity that can be found and referenced on the map: the edge of a curb, a street intersection, etc.). No matter how precise these measurements are, however, the location of the water main will not be accurate if these other reference points cannot be accurately located. And those reference points cannot be accurately located. And those reference points will not be accurately located if the locations of other reference points to which they are tied are not accurate—and so on. Somewhere, the accuracy buck has to stop so that everyone doing these measurements can rely on some reference points that are universally accepted as accurate.

Fortunately, there are physical reference points that can be used to map information accurately for the base map of an urban geographic information system. These physical reference points are referred to as *survey monuments,* or permanent manmade physical markers that are placed on the earth and from which accurate measurements (called *surveys*) can be made to locate physical and legal features on maps. When these monuments are used to map data or to combine data from two or more maps, they are then referred to as a *survey control network.*

The accuracy of maps, then, is a function of how well locations on maps can be found in relation to monuments (also called *survey control points* or *control points*) in a survey network. This means that a surveyor, using a map to find the physical location of a feature on the map, first locates a physical monument that has a previously recorded ("known") location and then uses the information on the map to move from the monument to the location represented on the map. How well the information on the map helps the surveyor find the entity (starting at the monument) determines the accuracy of the map. A clear, yet comprehensive, explanation of map accuracy, the issues surrounding it, and how it affects different applications in local government is presented by Croswell (1984), who gives the following definition of map accuracy:

> Horizontal map accuracy is simply defined as the conformity between the position of features plotted on a map and their true position defined relative to an accepted control network or datum. It is expressed in terms of an error, in linear units, between the position of the plotted map feature and its true position.   p. 50

While there are a number of accepted standards for horizontal map accuracy [most notably the National Map Accuracy Standards developed by the U.S. Bureau of the Budget in 1947, the U.S. Department of Transportation Map Standards (1968), and the Proposed Spatial Accuracy Standards for Large Scale Line Maps proposed by the American Society of Photogrammetry (Merchant, 1987)], determining the "true position" of a physical entity such as a monument can cause confusion because different maps are based upon different survey networks.

Survey Networks

A survey network (often called a *survey control network*) consists of permanent physical monuments whose locations are recorded and used by surveyors as a reference to locate features or to record the locations of features for future reference. If these physical monuments can be found on the earth and if their locations are accurately recorded, then all features, whether they are physical or not, can be found or recorded when accurate measurments are made from the monuments to which they are referenced. The accurate location of a manhole, for example, can be recorded by measuring its distance and bearing from a monument in a survey network. By using the recorded location of the monument and performing geometric calculations on the distance and bearing measurements to the manhole, the location of the manhole can be determined. Similarly, the corner of a parcel of land can be located (even though it cannot be seen) by first finding the survey network monument to which it is referenced and then measuring from it using distance and bearing measurements recorded for that parcel corner on the deed or other legal survey record.

Monuments, then, are reference points to which all measurements are made for locating features and recording their locations on maps in a survey network. Locations of monuments and features can be recorded in a number of different ways: coordinates from a local coordinate system; coordinates from a national coordinate system such as the State Plane Coordinate System; measurements from township, section, or quarter-section corners in the Public Land Survey System; or even measurements from observable physical features such as "the big oak tree at the intersection of Adnan Road and Pleier Drive." Local governments vary in the use of these reference systems, some using their own *local survey network*, whose monument locations are not related to other monuments across the country nor to the global surface of the earth, and some using a *geodetic survey network*, whose monument locations are referenced to the earth's surface (the geoid) and thus are related mathematically to all other locations on the surface of the earth.

Local Survey Networks

Most local governments use local survey networks for cadastral maps, site plans for development projects, as-built drawings for public works and utility construction projects, and other maps requiring precise surveys. In many cases the monuments used for taking measurements are site-specific and nonpermanent (a PK nail in a railroad tie, a wooden stake, a metal pipe) and therefore are subject to either movement or, worse, destruction. After they are used for the project, they may not be usable for future projects that may require a survey in the same area and therefore may have to be re-created from other features or not used at all. In these instances, the possibility for errors increases, which usually causes discrepancies in the surveys of adjacent projects. While the measurements made in surveys using local survey networks may be very precise and quite adequate for the specific project or need at the time, they are not accurate with respect to the earth's geodetic grid and, therefore, other features on the earth's surface.

Using precise measurements in a survey referencing a local survey network that is not tied to other features on the earth's surface results in what Croswell (1987) terms "the phenomenon of the 'floating plat' which is accurate within itself, but cannot be tied to an area-wide map base or other plats" (p. 53). When these plats are used to build the cadastral information for a base map in an urban geographic information system, adjacent property boundaries from previously mapped plats do not match, and it becomes necessary to adjust one or the other boundary lines ("fudge," as Croswell puts it) to make them match for a continuous map of the entire jurisdiction. These adjustments have a detrimental effect on the precision of the measurements in the plats.

Even when permanent monuments such as those used in the states that have been surveyed as part of the Public Land Survey System are used in a local survey network, there is no systematic method to match maps that are coincident (cadastral and planimetric, for example) or adjacent (such as neighboring plats), because when the monuments were created they were not related to the earth's geodetic grid.

PUBLIC LAND SURVEY SYSTEM

The Public Land Survey System (PLSS) is a common local survey network that is based upon the principles of property law and consists of a rectangular grid of monuments spaced 1 mile apart horizontally and vertically, defining an area of land called a "section." Established by the federal government immediately following the Revolutionary War as part of the Land Ordinance of 1785, the PLSS defines a location referencing system that is used by land surveyors in 30 of the 50 states.

The Land Ordinance of 1785 was one of the major government decisions of the Post-Revolutionary War period and had a profound impact on the nature of western settlement (Panel on a Multipurpose Cadastre, 1980, p. 9) because it addressed the following problems:

1. The United States had an enormous debt to France, Spain, and other countries for financial loans obtained during the Revolutionary War. Because Congress had no authority to levy taxes on lands, the sale of lands in the Northwest Territory was the best alternative for raising revenue in the shortest period of time (Hostrop, 1984, p. 253).

2. It satisfied the democracy advocates of the new government by subdividing the public land for sale into small parcels (sections) that could be afforded by the common people. This prevented control of the land by companies and wealthy men who could afford the higher prices associated with larger tracts (Hostrop, 1984, p. 253).

3. It established a systematic rectangular coordinate system based upon the 6-mile-square "township" as the basic survey unit. This corrected previous difficulties experienced with the "metes and bounds" descriptions of the land grants associated with property descriptions in the original 13 states. Such land grants often were made strictly on the basis of a written description without the benefit of a survey to determine property demarcations. In many cases, this lack of adequate surveys created problems that resulted in boundary disputes among property owners. The Land Ordinance of 1785, in creating the Public Land Survey System, provided for surveys to be made *prior* to the actual issuance of land grants and sales (Panel on a Multipurpose Cadastre, 1980, p. 9).

Other than additional provisions in 1804 to create "baselines" and "meridians" to control the locations of townships, the basic legislation regulating the PLSS has remained essentially unchanged since the American Revolution.

The rectangular coordinate system defined by the Public Land Survey System uses an area, the township, as its basic reference unit. Each township is subdivided into 36 1-mile-square sections that, in turn, are further subdivided into quarter-sections of 1/2 mile on a side in urban areas. In order to locate the townships, a grid of lines running north–south (range lines) and east–west (township lines) were established at 6-mile intervals. Thus, the rectangular system references a particular range line and township line in order to locate the township. Sections are numbered consecutively within a township, 1–36.

Range lines and township lines are referenced to a beginning point, defined in the PLSS as a "point of origin." When the survey was begun in 1785, the point of origin was established on the Ohio River where the states of Ohio and Pennsylvania meet. Since that time, 30 more points of origin have been established in the 48 contiguous states and 5 more in Alaska. These points of origin, as the first one in Ohio, are distinguished by natural features such as mountain peaks or the intersection of two rivers. From each point of origin baselines run due east and west, and meridians run due north and south. Range lines are set parallel to meridians and are numbered consecutively from each meridian. Township lines run parallel to baselines and are numbered consecutively from each baseline (see Figure 6.5).

The Public Land Survey System offers a simple means to identify a parcel of land ("Section Number 16 of Township Line 3 North and Range Line 2 East") compared with the metes and bounds descriptions

FIGURE 6.5. The rectangular coordinate system of the Public Land Survey System. Six-mile square townships are defined by township lines (T.1N., T.2N., etc.) running parallel to baselines (east-west) and rangelines (R1E., R.2E., etc.) running parallel to meridians (north-south).

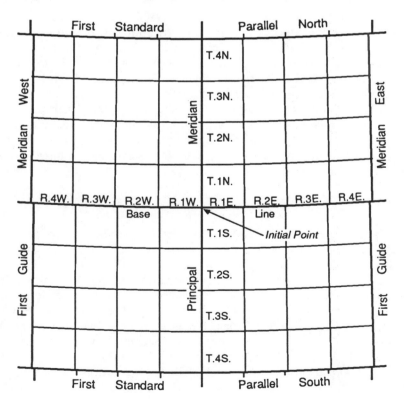

of distances, angles, and bearings from a physical feature. It also provides for physical monuments to be placed at section corners so that the section boundaries can be identified when property is conveyed or subdivided.

Since the original PLSS survey began over 200 years ago, the monuments at the section corners have become legally accepted reference points for further subdivision of the sections. When land became settled and smaller tracts were needed in densely populated areas, these small tracts of land were surveyed and referenced to section corner monuments. Subdivisions, for example, are recorded by a local government when their surveys reference section (or even quarter-section) monuments. Thus, not only can subdivisions be mapped on a section or quarter-section map, but all of the lots and blocks within the subdivision can also be mapped.

Unfortunately, most of the monuments set in the ground over 200 years ago have long since been obliterated by the elements of nature and the activities of humans (initially, wooden stakes or other nonpermanent markers were used as monuments). While their locations have been accurately recorded on maps (by reference to section numbers, range lines, and township lines), the physical monuments themselves often cannot be found. Accordingly, local and state governments have been conducting programs over the past 100 years to reestablish the monuments based upon their recorded locations in reference to monuments that have remained intact. These "remonumentation" programs, as one can expect, are time-consuming and expensive. For this reason, many PLSS monuments still remain obliterated, and, as a result, surveys must be tied to other PLSS monuments farther away. This not only requires more time to conduct a survey, but also introduces errors to the measurements when traditional surveying techniques are used. This results in adjacent section or quarter-section maps that do not fit together when they are digitized for establishing a base map for an entire jurisdiction.

Geodetic Survey Networks

Survey networks that are tied to the earth's geodetic grid provide absolute accuracy because their coordinates are global in nature and can accurately locate a feature that can be referenced to any other feature on the earth. No "floating plats" occur, and such maps can be easily integrated (overlaid) with other maps for the same geodetic area as long as the other maps are also registered on the same survey network.

One geodetic survey network that is in widespread use is the *National Geodetic Reference System* (NGRS), developed by the federal government. The National Geodetic Reference System is a nationwide network of 750,000 monuments whose precise geographic locations and elevations have been established and recorded by the National Geodetic Survey (NGS) of the National Ocean Service, National Oceanic and Atmospheric Administration (NOAA). The "triangulation and transverse stations" established by this federal agency are monumented points whose geodetic positions in terms of latitude and longitude are known with such accuracy and precision that any station, if its monument is destroyed, can be restored within very close tolerances to its original position by surveys initiated from other stations in the network. Since line-of-sight measurements are used, the curvature of the earth limits their spacing to about 10–20 miles apart (sometimes up to 75 miles in alpine terrain). Observations from one station to another are usually made from mountaintops or specially designed steel towers that are erected above the monuments and then dismantled after the measurements are taken.

Assigning coordinates to these stations requires accurate measurements that are based upon the principles of the science of geodesy—describing the shape of the earth so that its surface can be represented mathematically. Latitude and longitude are *spherical coordinates* that were devised nearly 2200 years ago to reference locations and measure distances over the entire three-dimensional globe of the earth. Latitude and longitude coordinates are recorded in degrees, minutes, and seconds, which are useful for large areas, but, for small areas such as countries and states within countries, they are cumbersome to use for computing distances. For this reason, *rectangular coordinates*, based upon a Cartesian system using XY coordinates, were devised for maps. Rectangular coordinates consist of two sets of straight, parallel grid lines that are equally spaced and perpendicular to each other. This allows the use of distance units in decimals, which are easier to use than degrees, minutes, and seconds.

In order to use rectangular coordinates, however, the curved surface of the earth must be transformed to a plane surface so that a two-dimensional map can be created from the earth's three-dimensional surface. This transformation is called a *map projection* and consists of mathematical computations on points defined on a three-dimensional shape that approximates the shape of the earth. The shape of the earth is called the *geoid*—a three-dimensional spheroid that coincides with the surface of the earth at sea level, extends (in an imaginary surface) through the continents, and has a direction everywhere perpendicular to the direction of gravity. Although the geoid appears to be a sphere, it is actually an ellipsoid that is flattened at the poles and bulges at the equator. The undulating surface of the geoid, however, makes the use of the geoid as a surface for mathematical computation impossible. Thus, an imaginary ellipsoid (which can be used for mathematical computations) is used to approximate the earth's surface when producing a map projection for defining rectangular coordinates.

A single ellipsoid defined mathematically for the entire globe suffices when it is desired to represent distances between features on the small scale that covers the entire earth. Since this ellipsoid only approximates the surface of the earth, however, its mathematical representation is not sufficiently accurate for measurements in smaller regions of the earth that are portrayed on large-scale maps. For example, the earth's surface in North America is more accurately represented mathematically by a portion of an ellipsoid that is different from the ellipsoid that most accurately represents the entire world (see Figure 6.6).

THE NORTH AMERICAN DATUM

An ellipsoid that accurately represents the surface of the earth in a specific region and has a mathematical representation allowing a point in the region to form an origin for coordinates on its surface is called a *local datum*. (An ellipsoid that approximates the entire globe and has its center at the earth's center of gravity is called a *global datum*.) The local datum used in North America is called the North American Datum and was first established in 1913. In 1927, a different ellipsoid was determined to be more accurate in approximating the earth's surface in the region and a new datum was adopted: the *North American Datum of 1927*, or NAD 27. This datum was generally accepted as the standard for almost 50 years when, through the use of satellites and other sophisticated technology, and the addition of Alaska to the region, a new datum was adopted by the federal governments of North America in

FIGURE 6.6. A *global datum* is a reference ellipsoid that provides a mathematical approximation of the surface of the earth at sea level (the geoid). A different ellipsoid (the North American Datum), providing a more accurate representation of the geoid in North America, is called a *local datum*.

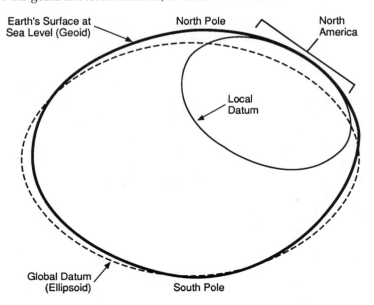

1986. This new datum, the *North American Datum of 1983*, or NAD 83, is a global datum that fits the entire geoid. This change from NAD 27 to NAD 83 has created considerable controversy among professionals in North America because all coordinates based upon NAD 27 have been changed to different coordinates based upon NAD 83. A detailed description of NAD 83 and its transformation from NAD 27 is provided by Morgan (1987).

MAP PROJECTIONS

Once the reference ellipsoid (a datum that provides a mathematical approximation of the earth) is established for a region, a map projection system is used to transform the three-dimensional surface of the ellipsoid to the two-dimensional surface of a map. It is impossible, however, to transform a sphere to a plane surface without modifying the geometric relationships (angles, areas, distances, and directions) of features on the surface of the sphere. While there are innumerable transformations that can retain one or several of these geometric relationships on a plane surface, there is no map projection system that can retain all of them. Consequently, there are over 250 different systems of map projections currently devised and described in the literature of mathematical cartography, each one emphasizing a particular combination of geometric relationships. The most widely used of these transformations in North America are: Lambert's conic projection, the Transverse Mercator projection, and the stereographic projection. These projections use the same reference ellipsoid, but they produce different reference grids (and their resulting maps) because each one emphasizes a different aspect of accuracy, depending upon how the maps are to be used.

A technical description of many map projection systems is provided by Robinson, Sale, and Morrison (1978), who summarize the characteristics of these three systems:

Lambert's conic projection—"Area deformation between and near the standard parallels is relatively small; thus the projection provides exceptionally good directional and shape relationships for an east–west latitudinal zone. Consequently, the projection is much used for air navigation in intermediate latitudes and for meteorological charts." (p. 63)

Transverse Mercator projection—"Since scale [distance] deformation increases away from the standard meridian, it is useful for only a small zone along that central line." (p. 62) "In recent years it has become popular for topographic maps and as a base for plane coordinate systems." (p. 63)

Stereographic projection—"The deformation (in this case, area exaggeration) increases outward from the central point symmetrically. This is an advantage when the area to be represented is more or less square or of continental proportions." (p. 63)*

Because urban geographic information systems use base maps that cover relatively small areas compared with the surface of the earth or even an entire continent, the choice of a map projection system to define coordinates for large-scale maps is limited to Lambert's conic projection or the Transverse Mercator projection. These two projection systems are used to establish the plane that provides rectangular coordinates for large-scale maps used by local governments.

Coordinate Systems

As stated previously, the coordinate systems used for base maps in urban geographic information systems can consist of the spherical coordinates of latitude and longitude and rectangular coordinates covering relatively small areas of the earth's surface. Rectangular coordinates, you will recall, are preferred over latitude and longitude because they are based upon a decimal system of measurement, which makes calculations much easier than using degrees, minutes, and seconds.

Universal Transverse Mercator System

While individual countries throughout the world have developed rectangular coordinate systems to suit their own needs, the *Universal Transverse Mercator* (UTM) grid system has been generally accepted worldwide for topographic maps, satellite imagery, and other small-scale mapping applications. The rectangular UTM coordinate system uses the Transverse Mercator map projection in an area of the entire world between 80 degrees North and 80 degrees South latitude into north–south columns 6 degrees wide called "zones." Each zone is divided into "quadrilaterals," which are 8 degrees high, forming a grid, each cell of which is referenced by a letter identifying its row and a number identifying its column. Each quadrilateral is divided into 100,000 meter zones designated by a system of letter combinations. Further subdivision of these zones is possible to the accuracy of zones of only 1 meter square. This accuracy (1 meter) is adequate for certain small-scale mapping applications, but local governments using an urban geographic information system use a different rectangular coordinate system: the *State Plane Coordinate System*, which is based upon the scientific measurements needed for the collection of earth science data for topographic, geologic, soils, and hydrographic maps.

*From *Elements of Cartography*, by A. H. Robinson, R. D. Sale, and J. L. Morrison, copyright © 1978, John Wiley & Sons, Inc. Reprinted by permission of John Wiley & Sons, Inc.

State Plane Coordinate System

The National Geodetic Reference System described earlier in this chapter is a survey control network that is used primarily for earth science data and is the basic control for all federal (and most private) topographic and other earth science mapping operations. In order to make the NGRS stations more readily available for local government use, the U.S. Coast and Geodetic Survey (now the National Geodetic Survey) devised the State Plane Coordinate System in 1933. This system translates the latitude and longitude coordinates of each station into rectangular coordinates on a plane surface (called a zone) for each state in the country (the larger states have more than one zone), each with its own projection, in order to improve the accuracy of the coordinates. Rather than project the reference ellipsoid used in North America, the North American Datum, onto a single plane surface covering the entire U.S., the State Plane Coordinate System projects the reference ellipsoid onto smaller planes (state planes). This limits the distortion on each plane to about 1 part in 10,000.

The State Plane Coordinate System permits engineers and surveyors in local government to tie their local surveys into the national geodetic network, thus eliminating the need to "fudge" coincident or adjacent features in a continuous base map. Thus, as Bauer (1984) states:

> The precise location on the earth's surface of all survey stations and landmarks established in local engineering and land surveys can be accurately described by stating their coordinates in reference to the common origin of the state plane coordinate grid.   p.135

In spite of this effort to make the NGRS stations more readily available to local governments, they are not generally used because the stations are located too far apart (10–20 miles) for practical use in local surveys that use optical equipment requiring line-of-sight measurements. Thus, only a very small percentage of land surveyors in the U.S. perform coordinate surveying with state plane coordinates; most use the traditional metes and bounds system by which properties are described on deeds and conveyances (Spradley, 1987, p. 185). Without further *densification* (the addition of more stations placed closer together), their use as reference points for state plane coordinates in an urban geographic information system is limited.

Global Positioning System

A new method available to land surveyors for coordinate surveying does not require monuments to be placed close together because it does not require line-of-sight measurements. The Global Positioning System (GPS) uses a constellation of (eventually) 21 satellites orbiting the earth at a very high altitude. The use of these satellites avoids the problems associated with land-based systems because line-of-sight measurements are eliminated and the monuments in the earth need not be replaced. All measurements are made from the satellites whose positions are known at all times.

Formally called the Navigation Satellite Timing and Ranging Global Positioning System (NAVSTAR GPS), GPS technology was developed by the U.S. Department of Defense to obtain instantaneous positions for worldwide, all-weather navigation. It has recently been made available for nonmilitary uses, allowing more accurate geodetic surveys by using specially designed receivers and more refined observing and data-processing techniques (Hanson, 1984, p. 8).

The basis for GPS is triangulation—measuring the distance a point on the earth is to three or more satellites simultaneously. Hurn (1989) explains the concept of triangulation with satellites, also called "satellite ranging" (see Figures 6.7, 6.8, and 6.9):

> By ranging from three satellites we can narrow down where we are to just two points in space. . . . How do we decide which one of those two points is our true location? Well, we could make a fourth measurement from another satellite. Or we can make an assumption. Usually, one of

FIGURE 6.7. A measurement from one satellite locates a point somewhere on a sphere with the satellite in the center. Source: Hurn, Jeff, Trimble Navigation, Ltd., *GPS A Guide to the Next Utility* (1989).

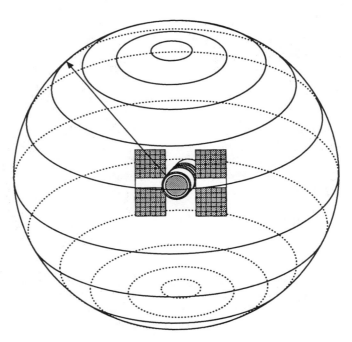

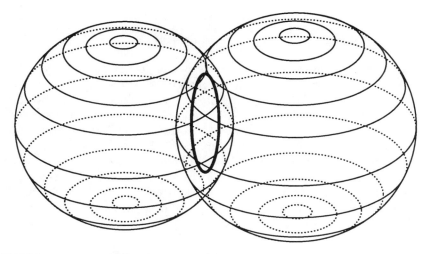

FIGURE 6.8. A second measurement from another satellite narrows the location of the point to any points on the circle where the two spheres intersect. Source: Hurn, Jeff, Trimble Navigation. Ltd., *GPS A Guide to the Next Utility* (1989).

FIGURE 6.9. A third measurement from yet another satellite puts the location of the point at one of two points where the third sphere intersects the circle. Source: Hurn, Jeff, Trimble Navigation, Ltd., *GPS A Guide to the Next Utility* (1989).

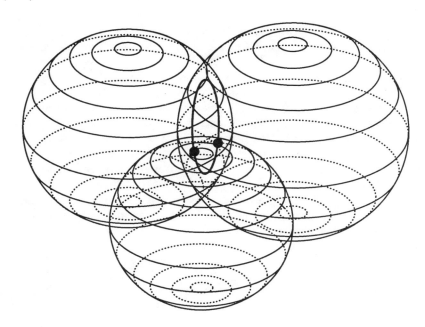

> the two points is a ridiculous answer. The incorrect point may not be close
> to the earth. . . . The computers in GPS receivers have various techniques
> for distinguishing the correct point from the incorrect one.   p. 16

The key, then, to measuring distances to satellites is the GPS geodetic
receiver, an electronic device that is placed on a feature on the earth
and receives signals from the satellites that are used to calculate the
distances. It consists of one or more radio channels to receive signals,
an internal clock, and a computer to perform calculations and store
information. Since some receivers are portable, they have their own
power source so that measurements can be taken in remote locations.

The receiver's clock is important because it measures how long it
takes for a signal to reach the receiver from the satellite. Once the travel
time of the radio signal is known, the distance can be computed from
the familiar equation:

$$distance = (velocity) \times (travel\ time).$$

Since radio waves travel at a velocity equal to the speed of light (186,000
miles per second), the GPS satellite receiver computes its distance from
a satellite by measuring how long it takes a radio signal to travel from
the satellite to the receiver.

The radio signal that is sent from a satellite consists of two pieces of
information: the time the signal was sent and a "pseudorandom code"
that uniquely identifies the signal. After the signal is received and clocked
by the receiver, it compares the time it was sent (the satellite's time) to
the time it was received (the receiver's time) to determine the travel
time of the pseudorandom code from satellite to receiver. This figure
is then multiplied times 186,000 to obtain the distance that the receiver
is from the satellite.

Obviously, both the satellite's clock and the receiver's clock must be
accurate and synchronized in order for this computation to yield accurate results (an error of only 1/100 of a second can produce a measurement error of 1,860 miles!). While each satellite has four atomic clocks
(three are available for backup) that are the most stable and accurate
time references ever developed and cost over $100,000 each, land sur-
veyors cannot afford receivers with such expensive clocks. Most receivers have inexpensive, yet very precise clocks that measure time with
nanosecond (10^{-9} second) accuracy. Even with such accuracy, however, a 1-nanosecond error in the travel time measurement can produce
a distance error of almost 1 foot. To avoid the expense of an atomic
clock in each receiver, then, a reading from a fourth satellite is used to
adjust the measurements based upon the inaccuracies in the receiver's
clock.

Since the GPS satellites are in such high orbits around the earth (about
11,000 miles), they are free of the earth's atmosphere and, like the moon,

reliably spin in this orbit according to simple mathematics. Because there is no atmospheric drag, the satellites orbit in a very predictable orbit that is known in advance. For this reason, the exact location of each satellite is known at all times, and some GPS receivers actually have this "almanac" programmed into their computer's memory so that they know where each satellite will be at any given time.

While GPS technology may eventually eliminate the need for additional densification of physical monuments on the earth and can produce faster and more accurate position readings than conventional ground survey methods, it will still be a number of years before the full benefit of GPS can be realized. Current restrictions which limit the full benefit of GPS include:

1. Not all of the planned 21 GPS satellites are currently in orbit around the earth. This restricts the "window" during which signals can be received and thus delays the response in obtaining positions.
2. The sophisticated technology of the GPS receivers has kept their cost above the cost of conventional first-order surveying instruments. As a result, not many surveyors or local governments have begun to use them.
3. There is no assurance that the satellites (or their replacements) will be available in the future, because, as with monuments, they must be maintained. Unforeseen political, economic, or other conditions could prevent this maintenance by the federal government.

Soon these obstacles will be overcome, and surveying with GPS technology will be routine. As improvements and other assurances are made and costs are amortized, surveyors and engineers will begin to accept GPS technology as a new convention for accurate survey control networks. As Hanson (1984) predicts:

> These and other improvements will virtually eliminate at least one of the obstacles to multipurpose land information systems—the lack of a readily accessible spatial reference system. It will no longer be necessary to extend geodetic networks by costly and time-consuming ground survey methods. Instead of creeping along the ground chaining one point to another with 10-mile links over a period of weeks, control networks will be extended in single 50-mile jumps in a matter of days. As a result, control for mapping will be reduced in expense from roughly half the cost of a mapping project as of today to a small fraction thereof.

> When fully perfected, GPS will revolutionize geodesy and rapidly supersede all current horizontal positioning methods. It will be faster, cheaper, and more accurate than anything in existence today or projected for the near future. Points can be located wherever needed. They will not have to be located on mountaintops, and no observing towers will be required, because they need not be visible from neighboring points. The receivers will be small, lightweight, and easily portable as backpacks. They will be au-

tomatic, that is, they can be set up, turned on, and left to receive and record signals for later processing at a central site. Observing times will be on the order of an hour or two, day or night, in almost any weather.

It may also be possible to reduce the number of costly, permanent boundary and control point monuments. Temporary monuments may be positioned by GPS as the need arises and removed when no longer needed, thereby cutting maintenance costs. Certainly, the numbers of markers employed for the purpose of extending control into new areas will decrease because of the longer jumps made possible by GPS.

It should be noted, however, that traditional ground surveying techniques will continue to be needed in areas that do not present a suitable window for receiving GPS signals. A good example of such an area in local government is in a central business district with many high-rise buildings that can obstruct the reception of direct line signals from the satellites.

Location Identifiers

Once an urban geographic information system has an established base map of cartographic information from cadastral or planimetric maps (or a combination of both), and a continuous coordinate system based upon a survey control network covering, at a minimum, the geographic area of the entire jurisdiction, it needs *location identifiers* to link the features represented on the maps to the nongraphics data describing those features. The concept of a location identifier is simple: It is a unique code (a number or a combination of numbers and letters) that not only is used as a record identifier in a nongraphics data base, but also represents a unique feature that can be identified on a map.

A given street light, for example, can be assigned a unique identifier that distinguishes it from all other street lights in the city. This identifier is used in a data base of descriptive information to distinguish the attributes of this street light from the attributes of the other street lights. A symbol representing the street light must also be displayed on a map so that the location of the street light can be distinguished from the locations of other street lights on the map. The identifier itself may be as simple as the address of the nearest property, or it may be as complicated as an *XY* coordinate pair. It also may be a completely random number unique to the entire jurisdiction. Whichever convention is used, the identifier must be *unique* (so that its attributes can be correctly associated with the feature it represents), and it must be *locatable* on a map (so that the symbol used on the map represents the specific feature the attributes describe). This provides the capability, for example, to count all street lights that have lamps older than 90 days in certain geographic areas.

Segment-based features such as water mains, sewer mains, gas lines, street pavements, etc. require the same standards (a unique identifier and locatable on a map) to be used in a geographic information system. Some jurisdictions use an address range to define a segment of the feature uniquely. Others use a beginning point (an address or an XY coordinate) plus the length of the segment. Still others use a unique code (such as a DIME record number) to identify the segment and place nodes where different segments meet. As long as the identifier can be linked to a line feature on a map, the segment-based feature can have its attributes geoprocessed by a geographic information system. Once water mains are linked in this manner, for example, a GIS can display all water mains that are older than 50 years or compute the total linear feet of all water mains that are older than 50 years and are in a certain district.

Area features such as land parcels, blocks, census tracts, and various administrative, political, and natural districts require the same standards for use in a GIS: unique identifiers and map references that uniquely identify them on a map. Thus, a census tract number that identifies a record in the data files provided by the Census Bureau must also be associated with the cartographic features that define the boundaries of the census tract on a map.

Land parcels, because they are relatively small geographic areas in relation to the entire jurisdiction of a local government, do not necessarily require their identifiers to be related to their boundaries. Many large jurisdictions, for example, do not create parcels as individual polygons, but choose rather to create the parcel boundaries as separate lines that, when displayed on a map, correctly portray the parcels. Parcel identifiers are then usually digitized on a separate overlay that can be combined with the parcel boundary overlay to display the parcels with their identifiers. Without explicit polygon definitions, these parcels cannot be manipulated as geometric entities (to perform area calculation, area shading, point-in-polygon determination, etc.). They can, however, be displayed and viewed by the map user for many valuable applications (parcel map retrieval, parcel attribute display, parcel map overlay with other geographic features, etc.) as long as unique parcel identifiers are included in the base map information. These identifiers should be located inside the parcel so that their XY coordinates can be used to associate attribute data with the correct location of the parcel when its boundaries are displayed.

Deciding on creating parcels as polygons or only as lines that meet visually can be important in an urban geographic information system, because polygonization of parcels is very expensive and time-consuming but can produce many more beneficial applications. The decision depends upon the results of an analysis of costs and benefits by the local government.

The Multipurpose Cadastre

Land parcels are an important component in an urban geographic information system because most of the data collected, maintained, and used by local governments are associated with properties and recorded by an address. Tax assessments, building permits, housing code inspections, fire responses, crime investigation, water service delivery, health inspections, and many more functions of local government record attribute data by property address. A geographic information system, then, that uses properties and their addresses as part of its base map can be used to integrate and aggregate data from these various sources of information for many service delivery, management, and policy analysis applications, as described in Chapter 3.

The need for accurate and current property records has received considerable attention among professionals who have seen their value when used in geographic information systems, especially at the local level. These professionals have observed, however, that current land records systems in the U.S. have many inherent problems that require attention before the full benefit of a GIS can be realized. Inaccuracies, lack of accessibility, duplication of data recording functions, inability to combine land records with other physical map features (such as the planimetric features recorded on the Census Bureau's TIGER maps and infrastructure records displaying the location of above- and below-ground public facilities), and in some cases, a lack of maps identifying property boundaries have led these professionals to call for a five-part program to improve local property records: the *Multipurpose Cadastre*.

The Panel on a Multipurpose Cadastre of the Committee on Geodesy of the Assembly of Mathematical and Physical Sciences of the National Research Council has explained the Multipurpose Cadastre in the Executive Summary of its report, *Need for a Multipurpose Cadastre* (1980):

> The concept of the multipurpose cadastre is a framework that supports continuous, readily available, and comprehensive land-related information at the parcel level. The components of a multipurpose cadastre are:
>
> 1. A reference frame consisting of a geodetic network;
> 2. A series of current, accurate large-scale maps;
> 3. A cadastral overlay delineating all cadastral parcels;
> 4. A unique identifying number assigned to each parcel that is used as a common index of all land records in information systems; and
> 5. A series of land data files, each including a parcel identifier for purposes of information retrieval and linking with information in other data files.*

Successfully implemented and combined with geographic information systems technology, the Multipurpose Cadastre can allow the in-

---

*Reprinted from "Need for a Multipurpose Cadastre," 1980, with permission from the National Academy Press, Washington D.C. (pp. 1–2).

tegration of virtually all land-related data into a common data base that can respond to the information needs of many local governments, utility companies, title insurance firms, real estate agents, land developers, and neighborhood organizations. While the information needs of these various organizations differ, a GIS can be built with enough flexibility to provide a series of overlays that can be combined, analyzed, and integrated with nongraphics attribute data to meet specific needs. Figure 6.10 is an example of how various map overlays can be integrated for a 1-square-mile section of a township in Wisconsin.

Some successful urban geographic information systems exist without parcel maps and parcel identifiers. In these systems, block outlines form the basic geographic unit to which geographic information is linked (Figure 2.9 in Chapter 2, for example, represents the association of nongraphics data, population by block, to maps containing block outlines). While property ownership and other property-specific data cannot be processed in these systems, a variety of small-scale analysis applications can be implemented. In these systems, when property data are needed in the analyses (such as the lead poisoning example in Chapter 3, in which housing age was related to age of population by census tract—see the section on Health Policy on Lead Poisoning), the process first requires data aggregation to a common geographic level (such as census tract) by geocoding with a DIME File, TIGER File, or other GBF, and then integrating the summary data horizontally across data sources (property records and census files).

Whatever the geographic level at which base map information is collected, stored, and maintained in an urban GIS, if the geographic features can be uniquely identified on a map that is registered on a continuous coordinate system based upon a geodetic control system, then geographic data and map integration are possible.

GIS County Exercise 3: Land Records Information

There are 88 properties in GIS County. The following maps (Figures E6.1, E6.2, E6.3, and E6.4) show their land use, parcel numbers, addresses, and parcel centroids.

The county tax assessor has assigned a unique parcel number to each property and keeps track of its land use, number of dwelling units, and year built. The tax assessor's list, a computerized file of this data, is shown in Figure E6.5.

Using the county's GBF, geocode this list to identify the block number and district number for each parcel (remember: odd numbers are in the left polygon and even numbers are in the right polygon). Then perform the five exercises.

Add the parcel centroids to the point table.

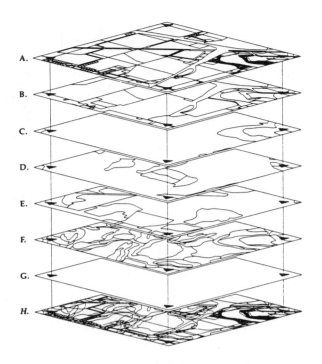

Concept for a
Multipurpose Land Information System

Section 22, T8N, R9E, Town of Westport, Dane County, Wisconsin

Data Layers:	Responsible Agency:
A. Parcels	Surveyor, Dane County Land Regulation and Records Department.
B. Zoning	Zoning Administrator, Dane County Land Regulation and Records Department.
C. Floodplains	Zoning Administrator, Dane County Land Regulation and Records Department.
D. Wetlands	Wisconsin Department of Natural Resources.
E. Land Cover	Dane County Land Conservation Committee.
F. Soils	United States Department of Agriculture, Soil Conservation Service.
G. Reference Framework	Public Land Survey System corners with geodetic coordinates.
H. Composite Overlay	*Layers integrated as needed, example shows parcels, soils and reference framework.*

FIGURE 6.10. In a Multipurpose Land Information System, map information is stored in overlays that can be combined to produce any desired composite map.

Illustration courtesy of Southeastern Wisconsin Regional Planning Commission (SEWRPC), Waukesha, Wisc. Reprinted by permission of the Wisconsin Land Records Committee (1987), ''Final Report of the Wisconsin Land Records Committee: Modernizing Wisconsin's Land Records, James L. Clapp (chair).

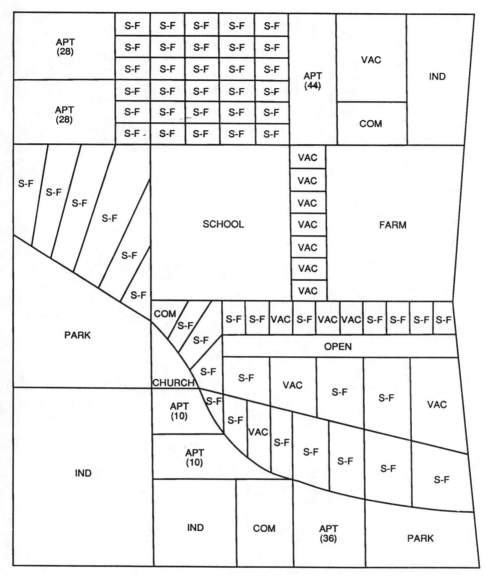

E6.1. GIS County land use map.

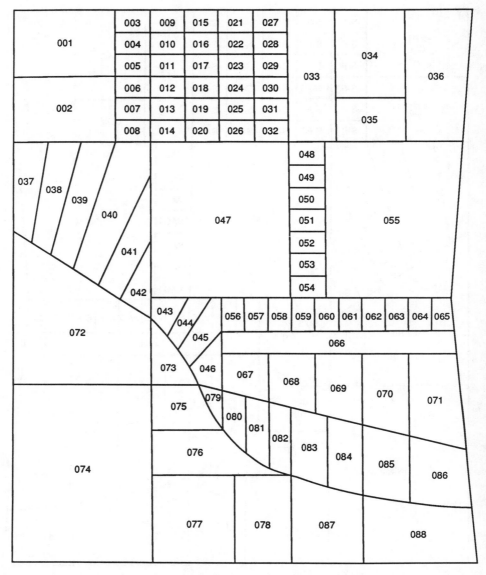

E6.2. GIS County parcel maps and parcel identifiers.

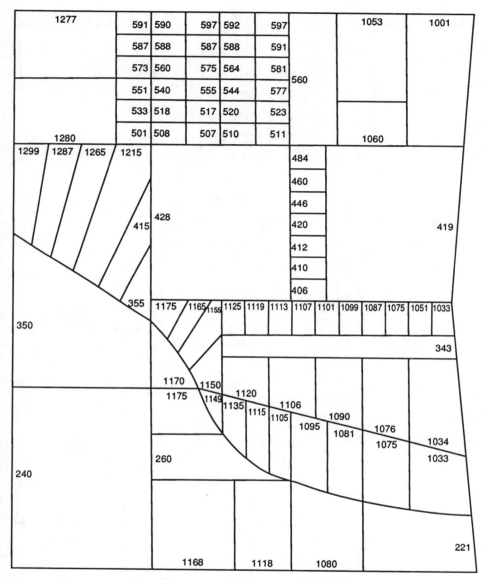

E6.3. GIS County parcel addresses.

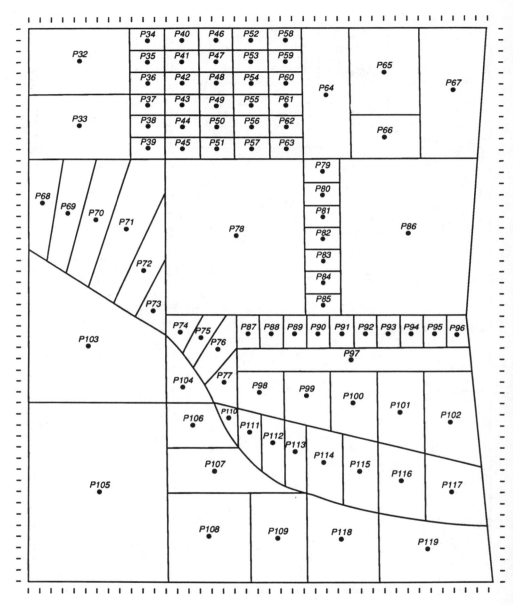

E6.4. GIS County parcel centroids.

218

PARCEL NO.	ADDRESS	LAND USE	DWELLING UNITS	YEAR BUILT
PARCEL 001	1277 W. ADNAN RD	APT	28	1954
PARCEL 002	1280 W. BECKIUS WY.	APT	28	1956
PARCEL 003	591 N. SADOWSKI ST	S-F	1	1945
PARCEL 004	587 N. SADOWSKI ST	S-F	1	1947
PARCEL 005	573 N. SADOWSKI ST	S-F	1	1944
PARCEL 006	551 N. SADOWSKI ST	S-F	1	1946
PARCEL 007	533 N. SADOWSKI ST	S-F	1	1945
PARCEL 008	501 N. SADOWSKI ST	S-F	1	1946
PARCEL 009	590 N. SADOWSKI ST	S-F	1	1926
PARCEL 010	588 N. SADOWSKI ST	S-F	1	1926
PARCEL 011	560 N. SADOWSKI ST	S-F	1	1927
PARCEL 012	540 N. SADOWSKI ST	S-F	1	1928
PARCEL 013	518 N. SADOWSKI ST	S-F	1	1929
PARCEL 014	508 N. SADOWSKI ST	S-F	1	1931
PARCEL 015	597 N. TIERNEY PL	S-F	1	1930
PARCEL 016	587 N. TIERNEY PL	S-F	1	1928
PARCEL 017	575 N. TIERNEY PL	S-F	1	1929
PARCEL 018	555 N. TIERNEY PL	S-F	1	1931
PARCEL 019	517 N. TIERNEY PL	S-F	1	1930
PARCEL 020	507 N. TIERNEY PL	S-F	1	1928
PARCEL 021	592 N. TIERNEY PL	S-F	1	1931
PARCEL 022	588 N. TIERNEY PL	S-F	1	1928
PARCEL 023	564 N. TIERNEY PL	S-F	1	1944
PARCEL 024	544 N. TIERNEY PL	S-F	1	1947
PARCEL 025	520 N. TIERNEY PL	S-F	1	1944
PARCEL 026	510 N. TIERNEY PL	S-F	1	1946
PARCEL 027	597 N. WINTER CT	S-F	1	1947
PARCEL 028	591 N. WINTER CT	S-F	1	1946
PARCEL 029	581 N. WINTER CT	S-F	1	1949
PARCEL 030	577 N. WINTER CT	S-F	1	1947
PARCEL 031	523 N. WINTER CT	S-F	1	1948
PARCEL 032	511 N. WINTER CT	S-F	1	1947
PARCEL 033	560 N. WINTER CT	APT	44	1952
PARCEL 034	1053 W. ADNAN RD	VAC	0	0
PARCEL 035	1060 W. BECKIUS WY	COM	0	1956
PARCEL 036	1001 W. ADNAN RD	IND	0	1959
PARCEL 037	1299 W. BECKIUS WY	S-F	1	1970
PARCEL 038	1287 W. BECKIUS WY	S-F	1	1946
PARCEL 039	1265 W. BECKIUS WY	S-F	1	1951
PARCEL 040	1215 W. BECKIUS WY	S-F	1	1881
PARCEL 041	415 N. SADOWSKI ST	S-F	1	1941
PARCEL 042	355 N. SADOWSKI ST	S-F	1	1950
PARCEL 043	1175 W. BERTRAND DR	COM	0	1962
PARCEL 044	1165 W. BERTRAND DR	S-F	1	1964
PARCEL 045	1155 W. BERTRAND DR	S-F	1	1964

E6.5. GIS County tax assessment list.

PARCEL NO.	ADDRESS	LAND USE	DWELLING UNITS	YEAR BUILT
PARCEL 046	1150 W. FROH ST	S-F	1	1964
PARCEL 047	428 N. SADOWSKI ST	SCH	0	1952
PARCEL 048	484 N. WINTER CT	VAC	0	0
PARCEL 049	460 N. WINTER CT	VAC	0	0
PARCEL 050	446 N. WINTER CT	VAC	0	0
PARCEL 051	420 N. WINTER CT	VAC	0	0
PARCEL 052	412 N. WINTER CT	VAC	0	0
PARCEL 053	410 N. WINTER CT	VAC	0	0
PARCEL 054	406 N. WINTER CT	VAC	0	0
PARCEL 055	419 N. VANDERHEYDEN WY	FARM	1	1876
PARCEL 056	1125 W. BERTRAND DR	S-F	1	1978
PARCEL 057	1119 W. BERTRAND DR	S-F	1	1980
PARCEL 058	1113 W. BERTRAND DR	VAC	0	0
PARCEL 059	1107 W. BERTRAND DR	S-F	1	1982
PARCEL 060	1101 W. BERTRAND DR	VAC	0	0
PARCEL 061	1099 W. BERTRAND DR	VAC	0	0
PARCEL 062	1087 W. BERTRAND DR	S-F	1	1980
PARCEL 063	1075 W. BERTRAND DR	S-F	1	1981
PARCEL 064	1051 W. BERTRAND DR	S-F	1	1984
PARCEL 065	1033 W. BERTRAND DR	S-F	1	1933
PARCEL 066	343 N. VANDERHEYDEN WY	OPEN	0	0
PARCEL 067	1120 W. FROH ST	S-F	1	1987
PARCEL 068	1106 W. FROH ST	VAC	0	0
PARCEL 069	1090 W. FROH ST	S-F	1	1987
PARCEL 070	1076 W. FROH ST	S-F	1	1988
PARCEL 071	1034 W. FROH ST	VAC	0	0
PARCEL 072	350 N. PLEIER DR	PARK	0	0
PARCEL 073	1170 W. DADLEZ DR	CH	0	1954
PARCEL 074	240 N. PLEIER DR	IND	0	1988
PARCEL 075	1175 W. DADLEZ DR	APT	10	1980
PARCEL 076	260 N. SADOWSKI ST	APT	10	1983
PARCEL 077	1168 W. KAPPEL RD	IND	0	1973
PARCEL 078	1118 W. KAPPEL RD	COM	0	1970
PARCEL 079	1149 W. FROH ST	S-F	1	1972
PARCEL 080	1135 W. FROH ST	S-F	1	1972
PARCEL 081	1115 W. FROH ST	VAC	0	0
PARCEL 082	1105 W. FROH ST	S-F	1	1974
PARCEL 083	1095 W. FROH ST	S-F	1	1978
PARCEL 084	1081 W. FROH ST	S-F	1	1975
PARCEL 085	1075 W. FROH ST	S-F	1	1981
PARCEL 086	1033 W. FROH ST	S-F	1	1979
PARCEL 087	1080 W. KAPPEL RD	APT	36	1971
PARCEL 088	221 N. VANDERHEYDEN WY	PARK	0	0

E6.5. GIS County tax assessment list (*cont'd*).

1. The county supervisors want to know the land use mix of their districts. Provide them with the following report:

GIS County Land Use Report

Dist	Land use	Housing units	Number of properties	Percent of district	Percent of county
A	Res				
	Nonres				
	Public				
	Vacant	_____	_____	_____	_____
	Total:				
B	Res				
	Nonres				
	Public				
	Vacant	_____	_____	_____	_____
	Total:				
C	Res				
	Nonres				
	Public				
	Vacant	_____	_____	_____	_____
	Total:				
D	Res				
	Nonres				
	Public				
	Vacant	_____	_____	_____	_____
	Total:				
	Res				
	Nonres				
	Public				
E	Vacant	_____	_____	_____	_____
	Total:				
	County totals:	_____	_____		

(*Note:* Res includes S-F and Apt; Nonres includes Com, Ind, and Farm; Public includes School, Church, and Park; Vacant includes Vac and Open.)

2. The U.S. Census Bureau has provided a computer tape of 1988 population data by block from the recently completed special census. The tape has the following information on it:

GIS County 1988 Population

Block Number	Total Pop.	Males <18	Males 18–65	Males >65	Females <18	Females 18–65	Females >65
01	130	12	37	18	17	25	21
02	36	5	5	7	7	5	7
03	40	2	3	10	10	4	11
04	89	11	14	19	10	14	21
05	18	4	6	0	2	6	0
06	0	0	0	0	0	0	0
07	1	0	0	0	0	0	1
08	0	0	0	0	0	0	0
09	0	0	0	0	0	0	0
10	39	6	10	3	8	10	2
11	0	0	0	0	0	0	0
12	20	0	2	8	0	1	9
13	9	3	3	0	0	3	0
14	12	1	4	0	3	4	0
15	72	24	11	1	15	20	1

a. Compute population summaries for each district, where residential density = population / (housing units):

GIS County Population by District, 1988

District	Population	Percent of County	Residential Density
A			
B			
C			
D			
E	————	————	————
GIS County total:			

b. Categorize each district by residential density according to the following ranges:

Category	Density Range	District	Category
I	less than 1.0	A	—
II	1.0–1.9	B	—
III	2.0–2.9	C	—
IV	3.0 and greater	D	—
		E	—

3. The Building Inspection Department reports that six buildings were built during the past month. It issues the following report to the Tax Assessor, Planning Director, County Supervisors, and the Media:

GIS County New Construction Month: January

Permit Number	Address	Building Type	Dwelling Units	Construction Date
10123	1036 W. Froh Street	Grocery	0	01/09/89
11003	1113 W. Bertrand Dr	Sin-Fam	1	01/18/89
11019	1055 West Adnan Road	Apartment	50	01/22/89
10268	1099 W. Bertrand Dr.	Sin-Fam	1	01/23/89
10010	1115 West Froh Str.	Sin-Fam	1	01/27/89
10125	1100 W. Froh St.	Sin-Fam	1	01/30/89

Update the following records:
   a. Tax assessor's list;
   b. County Supervisor's land use report.

4. The farmer at 419 N. Vanderheyden Way has sold her land to a local developer who is planning to build apartments and single-family homes on the land. The surveyor has submitted the plat shown in Figure E6.6 to the County Engineer for certification.

Using the maps in Figures E6.7 and E6.8, update the following records:
   a. Point, line, area tables;
   b. GIS County GBF;
   c. Tax assessor's list;
   d. County Supervisor's land use report.

5. The GIS County Office of Emergency Preparedness has received a map of official flood plains for the county. The flood plain map (shown in Figure E6.9) will be used by the Building Inspection Department to restrict major building construction and renovation within the flood plain. It will also be used to identify all property owners within the flood plain so that information about federal insurance programs can be disseminated. Also, the Office of Emergency Preparedness will use this map to help plan for the relocation of residents in the event that major flooding does occur. The Office, then, must integrate the information from the flood plain map with other property-related data in the county's GIS. Perform the following tasks:

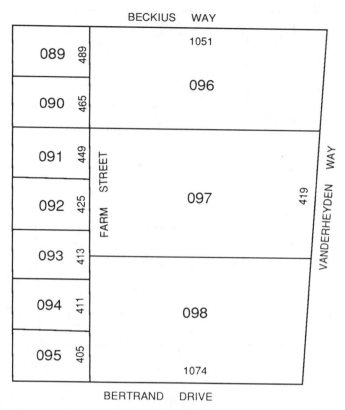

E6.6. GIS County land subdivision with new parcel identifiers and addresses.

a. Create a new polygon called "flood plain" and update the point, line, and area tables.
b. Using this new polygon, list all addresses inside the flood plain so that the Building Inspection Department can check all incoming building permit applications. This same list will be used to send the owners information about flood insurance.
c. Prepare a report summarizing the number of housing units that are within the flood plain. Break these totals down by district so that the county supervisors will know how their districts would be affected by a major flood.
d. Determine the percentage of each district (in area) that is in the flood plain.

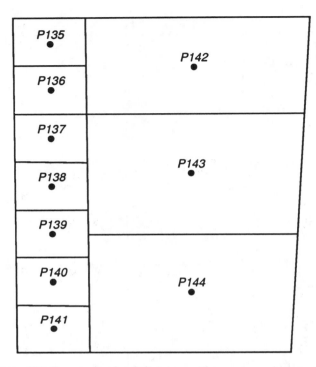

E6.7. GIS County land subdivision with new parcel centroids.

E6.8. GIS County land subdivision with new geographic identifiers.

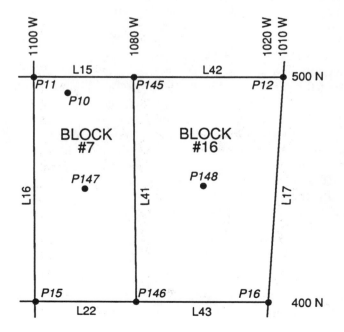

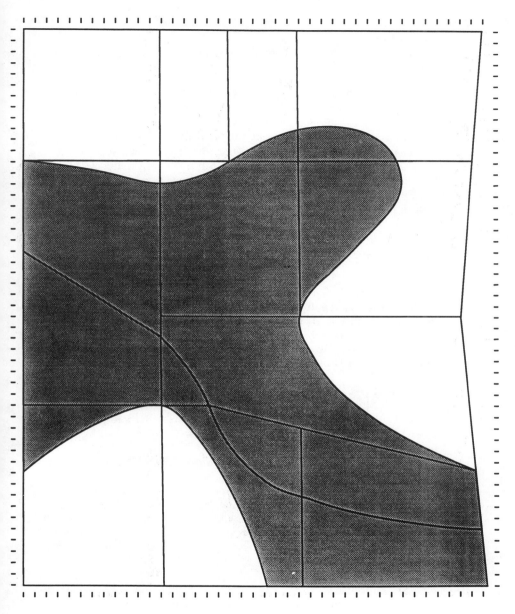

E6.9. GIS County flood plain map.

References

Ayers, Eunice H. (1984), "Implementation of Modern Land Records Systems: Politics and Institutional Reform," *Seminar on the Multipurpose Cadastre: Modernizing Land Information Systems in North America*, Bernard J. Niemann, Jr. (ed.), Wisconsin Land Information Reports: Number 1, Institute for Environmental Studies, University of Wisconsin—Madison, Wisc.

Bauer, Kurt W. (1984), "Public Planning and Engineering: the Role of Maps and the Geodetic Base," *Seminar on the Multipurpose Cadastre: Modernizing Land Information Systems in North America*, Bernard J. Niemann, Jr. (ed.), Wisconsin Land Information Reports: Number 1, University of Wisconsin—Madison, Madison, Wisc.

Croswell, Peter L. (1984), "Map Accuracy: What is It, Who Needs It, and How Much is Enough," *Seminar on the Multipurpose Cadastre: Modernizing Land Information Systems in North America*, Bernard J. Niemann, Jr. (ed.), Wisconsin Land Information Reports: Number 1, Institute for Environmental Studies, University of Wisconsin—Madison, Wisc.

Hanson, Robert H. (1984), "Advanced Positioning Technology: Implications for Multipurpose Information Systems," *Seminar on the Multipurpose Cadastre: Modernizing Land Information Systems in North America*, Bernard J. Niemann, Jr. (ed.) Wisconsin Land Information Reports: Number 1, University of Wisconsin—Madison, Madison, WI.

Holloway, Donald P. (1983), "North Carolina's Land Records Management Program: Progress Toward County Cadastres," papers from the annual conference of the Urban and Regional Information Systems Association, Washington, D.C.

Hostrop, Bernard (1984), "The Public Land Survey System (PLSS): The Foundation for a Multipurpose Cadastre," *Seminar on the Multipurpose Cadastre: Modernizing Land Information Systems in North America*, Bernard J. Niemann, Jr. (ed.) Wisconsin Land Information Reports: Number 1, University of Wisconsin—Madison, Madison, Wisc.

Hurn, Jeff (1989), *GPS A Guide to the Next Utility*, Trimble Navigation Ltd., Sunnyvale, Calif.

Merchant, Dean (1987), "Spatial Accuracy Specification for Large Scale Topographic Maps," Technical Papers of the 1987 ASPRS-ACSM Annual Convention, Volume 2. American Society of Photogrammetry and Remote Sensing, American Congress on Surveying and Mapping, Bethesda, MD.

Morgan, J. G. (1987), "The North American Datum of 1983," *NCGA'S Mapping & Geographic Information Systems '87 Proceedings*, National Computer Graphics Association, Washington, D.C.

Panel on a Multipurpose Cadastre, Committee on Geodesy, Assembly of Mathematical and Physical Sciences, National Research Council (1980), *Need for a Multipurpose Cadastre*, National Academy Press, Washington, D.C.

Robinson, Arthur, Sale, Randell, and Morrison, Joel (1978), *Elements of Cartography*, John Wiley & Sons, Inc., New York.

Spradley, L. Harold (1987), "The Use of Ground Truth Surveys in Performing NAD (83) Ties for Geographical Information Systems," *NCGA'S Mapping & Geographic Information Systems '87 Proceedings*, National Computer Graphics Association, Washington, D.C.

U.S. Bureau of the Budget (1947), "U.S. National Map Accuracy Standards," U.S. Bureau of the Budget, Washington, D.C.

U.S. Department of Transportation (1968), "Reference Guide Outline: Specifications for Aerial Surveys and Mapping by Photogrammetric Methods for Highways," prepared by the American Society of Photogrammetry, Bethesda, MD.

ADDITIONAL READINGS

Epstein, Earl F., and Duchesneau, Thomas D. (1984), *The Use and Value of a Geodetic Reference System*, Federal Geodetic Control Committee, University of Maine at Orono, Maine.

7

The Model Urban GIS Project

Ken [Brelsford], Johnson County, Kansas, was briefing officials on
the merits of GIS in city government. After politely and passively
listening for most of the presentation, the significance of what
[Brelsford] was saying finally penetrated enough to cause a response
from one of them, who leaned back in his chair and said, "What
are you trying to do, change the way we operate?" Said [Brelsford]
without a pause, "Yes, sir." GISing can be a lonely pursuit . . .

GIS World, May 1989

Officials in local government who are interested in implementing a
geographic information system have two important tasks ahead of them
to ensure success of the system: First, they must convince the decision-
makers to provide funds for the project; and second, they must deliver
on the promises they made during the decision-making process. Each
task requires a structured process that is consistent with a comprehensive plan involving many different resources. The process, however,
varies in complexity depending upon the complexity of the envisioned
system.

For low-cost, single-use, microcomputer-based geographic information systems that are used to perform spatial analysis of small-scale
data (census tracts, ZIP Code areas, aldermanic districts), the process
can be relatively simple. A variety of "desktop mapping" systems are
available from the private sector with prices ranging from a few hundred
dollars (if you already have a microcomputer) to less than $25,000 (in-
cluding the base map of the jurisdiction). Such small installations can
provide valuable tools for the many policy analysis and planning activities of local government. Often, these systems can be cost justified by
the mere fact that they can eliminate the need to manually cut and stick
ZIPATONE area shading symbols on choroplethic maps for reports and
presentations. Analyses of geographic data requiring a map accuracy
of more than 20 feet (such as determining fire response zones for fire
stations) are well served by these low-cost desktop mapping systems
and do not require extensive investments in hardware, software, and
support staff. Usually, existing staff can be trained and assigned responsibility for supporting the geographic information needs of a small

municipality or single department, but to ensure success this responsibility is best implemented when it is assigned to a single staff member or function (such as an "information center") where GIS expertise can be centralized.

Large multipurpose, parcel-based, geographic information systems that process and produce engineering-quality map information and contain flexible data structures that integrate data from various sources, however, have caused many GIS advocates to fail on either the convincing or delivering tasks. Because of the large time and dollar investments required for these systems, some professionals have experienced lifelong crusades trying to convince decision-makers that a GIS project is worthwhile. Those who have succeeded in obtaining the millions of dollars needed to implement these systems face the additional danger of failing to deliver the benefits originally anticipated in a reasonable time frame considering the cyclical nature of leadership in local government.

The issues associated with justifying and managing major GIS projects have drawn the interest of many professionals throughout the years. Managers of successful projects, consultants who have experienced a number of different efforts, and academicians who have studied the literature and surveyed the field have all identified a number of critical success factors that can make or break the effort to bring GIS technology to local government. These issues can be generalized into the following categories:

- *Evaluating geographic information needs*—Developing long-range GIS plans that are consistent with the goals of the organization and determining the geographic information needed to achieve those goals.

- *Gaining organizational support*—Performing a cost/benefit study to determine the economi~ value of a GIS and conducting a pilot project to test various app   .tions and refine the cost and benefit estimates.

- *Managing the GIS project*—Converting the map information to digital form and organizing resources for continued support and future expansion.

It is important to address all of these issues in a consistent and professional manner because the investment in a geographic information system is considerable and, therefore, attracts a significant amount of attention when it is compared with other programs that compete for scarce tax dollars. As with most resources, the geographic data base and system that processes it also require constant maintenance and management once the system is implemented, and thus the decision to implement a GIS also includes a commitment to provide continuing operational support for the system.

Evaluating Geographic Information Needs

A complete understanding of how the organization uses geographic information in performing its various functions is necessary in order to identify the potential benefits of a GIS and ensure that its use will be consistent with the long-term goals of the organization. This effort is most valuable when the analysis includes all aspects of the organization so that information-sharing opportunities can be investigated and so that future expansion in later phases of the project can be properly planned.

The Long-Range GIS Plan

A comprehensive long-range plan that analyzes the needs of the organization over a 5–10 year time period will ensure that the use of the GIS will be consistent with the goals of the organization and thus prevent unrealized expectations and disappointment. The long-range plan allows decision-makers to evaluate the applicability of the system to the strategic and tactical plans of the city or county, and it ensures that the appropriate resources are available at the time they are needed during the course of its development.

A good long-range plan meets the following objectives:

1. Obtain high-level support—Not only does the long-range plan bring opportunities for improving government to the attention of decision-makers, but it also gives them confidence that those who advocate the new GIS technology are competent and can make the project succeed.
2. Identify all potential applications—A comprehensive long-range plan considers all of the geographic information needs of the organization and thus ensures that no improvement opportunities are omitted from consideration. This is especially important with geographic information systems because many local government officials are not aware of the full capabilities of the technology and may tend to assume that their needs cannot be satisfied by using a GIS.
3. Prioritize applications for orderly implementation—If a long-range plan is successful at identifying all potential applications and is related to the strategic and tactical plans of the city or county as a whole, then it is possible to schedule or prioritize the implementation of applications in the order that will be most beneficial to the organization. It also ensures that those applications or data bases that are needed to be in place prior to others that need them are, in fact, completed prior to continuing development or expansion.
4. Obtain maximum benefits organization-wide—Since one of the

major features of a GIS is its ability to integrate information from a number of different sources, the long-range plan can identify information-sharing opportunities that had never before been known, or had never been possible without a GIS. Separate functions that had never shared information before can realize improvements through geographic data integration. This assures a city- or county-wide benefit analysis and prevents specialized interests from impeding full benefit realization.

5. Identify resource requirements—One of the worst mistakes information systems specialists can make in advocating major development efforts using computers is to surprise decision-makers later in the life of the project with the need for additional unanticipated funds in order to complete the project successfully so that full benefits can be realized. Since most elected officials and other decision-makers are not familiar with information systems technology, they must trust their technical staff to present a comprehensive and understandable proposal upon which they can make a funding decision. When costs escalate afterwards, the credibility of those in charge is weakened. A long-range plan, one that is honest and objective in projecting costs, can prevent later surprises that may jeopardize full implementation because funding is not sufficient.

Above all, a long-range plan for a geographic information system must relate directly to the long-range plans of the local government it serves.

There are a number of generally accepted information systems planning methodologies that have been used for a number of years in the information systems industry to ensure that new technology will successfully support the long-term goals of an organization. The most popular of these planning methodologies, business systems planning (BSP), critical success factors (CSF), and the Nolan–Norton stages theory, all attempt first to define the functions and other factors that are most important to an organization and then develop the appropriate information systems strategies that will be most beneficial in supporting them. This process ensures that technology is applied to the most important functions and therefore will have the most impact on the success of the organization. Long-range planning for geographic information systems should follow the same approach.

RECORDING THE GOALS OF LOCAL GOVERNMENT

Some local governments have a long-range planning function that documents the direction in which the current administration desires to lead the jurisdiction in the future. Such long-range plans are quite general in nature, focusing on trends and problems requiring long-term programs for solution. The goals stated in such plans consist of: increasing jobs, improving the economy, providing more affordable housing, reducing crime, improving the health of citizens, improving the environment, creating more efficiency in government, improving education, maintaining the urban infrastructure, controlling growth, improving the quality of neighborhoods, and a host of other local issues that affect the taxpayer and the evaluation of leadership in government. If these goals are explicitly documented, the GIS advocate must use them in developing the long-range GIS plan. If they are not explicitly stated, there are surrogate sources of documentation available, the most available of which is the annual budget and all of its associated documents. Interviews with elected officials, their administrative assistants, and department heads can also substitute for explicitly stated goals.

IDENTIFYING LOCAL GOVERNMENT FUNCTIONS

All functions that contribute to the goals of the organization must be reviewed to determine how a geographic information system can be used to assist in their improvement. These functions, or tasks, are usually defined in the mission statements of the various departments, bureaus, divisions, sections, or other offices within the organizational structure of the government. Each one can be associated with a goal, otherwise it would not be performed. For example, the function "repair city streets" is a function usually assigned to a bureau within the Public Works Department of a local government. This function relates directly to the goal of "maintaining the urban infrastructure." Other functions also relate to this goal: "prepare the capital improvements program," "inspect, repair, prune, and maintain trees," "repair sewers, catch basins, storm inlets, and manholes," "repair, maintain, and replace water distribution facilities," and other such government functions that must be performed in order to maintain the infrastructure. All of these functions have a geographic component because they describe work by the government at specific locations within the jurisdiction. Table 7.1 is a list of urban functions that are representative of those in local government performed in order to "maintain the urban infrastructure."

Table 7.1. Government functions supporting the goal of "maintaining the urban infrastructure."

Prepare the capital improvement program and budget	Examine sewers to determine repair needs
Review development proposals	Repair sewers, catch basins, storm inlets, and manholes
Undertake redevelopment activities	Plan, design, and construct water service facilities
Administer public works contracts	
Bill for services and damage to government facilities	Locate and mark underground water service facilities
Perform research and planning for development needs	Modernize outdoor warning siren system
Respond to citizen requests for service and information	Test city water

Plan for emergencies
Administer permit applications
Administer, bill, and collect special assessment charges
Inspect work sites and investigate accidents
Plan paving program
Process Digger's Hotline requests
Provide information and maps for grant submittals
Prepare plans for pavement construction
Maintain as-built records for traffic lights and street lights
Inspect private sewer, water, street, and alley facilities
Supervise, coordinate, and inspect public facility construction
Maintain bridges
Analyze traffic accidents and recommend improvements
Investigate and report traffic accidents
Construct and repair electrical service to government
Inventory trees and nursery stock
Monitor and inspect Cable TV construction and pole attachments
Maintain and repair public parks and recreation facilities
Patch pot-holes, cave-ins, and defective sidewalks
Assist community groups undertaking facilities rehabilitation
Install and maintain police and fire communication facilities
Install and maintain traffic, crosswalk, and special markings
Prepare and maintain maps of public ways
Provide service and information for flood and sewer problems
Clean sewers, catch basins, storm inlets, and drainage channels
Inspect, maintain, prune, and replace trees
Remove trees and stumps
Maintain boulevards
Conduct tree-related research
Investigate hazards and poor sanitation in public facilities
Prepare plans for changes to the street lighting system
Install, operate, and maintain street and alley lights

Maintain and repair harbor facilities
Acquire and develop waterside sites for the Port Authority
Design traffic control facilities
Install, operate, and maintain traffic signals
Install, replace, and repair traffic signs
Construct new public parking lots
Plan and manage the repair and construction of bridges
Operate bridges
Remove ice and snow from streets
Operate recycling stations
Cut weeds in the public right-of-way
Remove and replace defective sidewalks
Remove and replace defective safety islands
Remove and replace defective handicap ramps
Remove and replace defective bus loading zones
Provide temporary street barricading service
Bill and collect water and sewer service charges
Maintain and repair water plant and system facilities
Install water mains, taps, and branch connections
Repair, maintain, and replace water distribution facilities
Inform the public on the locations of emergency facilities
Maintain and repair bridges and viaducts
Determine remedial work to prevent backwater and surface flooding
Maintain sewer system records
Plan for adequate sewer service
Sweep streets and alleys
Repair street and alley pavements
Test, repair, and exchange water meters
Read water meters
Purify and pump water to utility customers
Schedule vehicular equipment and drivers
Dispatch vehicular equipment and drivers
Dispose of abandoned vehicles towed from city streets
Dispose of toxic and hazardous substances

IDENTIFYING FACILITIES AND ENTITIES

Once the goals of the government have been set and the functions supporting those goals identified, the GIS planners must then identify which facilities and entities are the objects of those functions. Facilities and entities are the physical, legal, and other objects located throughout the jurisdiction either by physical placement (such as a house) or by cartographic representation on a map (such as a neighborhood). A function of urban government ("responding to fires") is always an action taken on some object (a building or a parcel of land). If that object has a geographic description that is important (the address of the building or the fire response zone for the fire station), then the GIS planners must consider the function as a potential for improvement through GIS technology. Obviously, some important functions of government are performed on facilities and entities that are not related to its geography, such as the function of "preparing payroll checks," which is performed on information about an employee (an entity) to produce a paycheck (another entity). Unless it is important to know *where* the employee worked, then the GIS planners can omit this function from their analysis.

Chapter 1 reported that 80–90 percent of all the information used by local government is related to geography. This means that the GIS planners must review almost all of the important functions of government to identify the facilities and entities that might be represented in its geographic information system. Infrastructure, housing, crime, public health, economic development, and many other important local issues are related to geography. The function of repairing city streets, for example, is performed on the facility called a "street," one of many infrastructure facilities (see Table 7.2). The street and alley facilities are composed of entities such as "pavement," "curbs," and "gutters," which are the components of the facility about which information is recorded. A long-range GIS plan should consider the importance of all the information relating to the facilities and entities that are objects of its vital functions.

Table 7.2. Facilities used in the function "repair city streets."

Urban infrastructure facilities	
Boulevard landscaping facilities	Sidewalks
Bridges	Street trees
Communications facilities	Street and alley lighting
Sewer system	Streets and alleys
	Traffic control facilities
	Water system

DETERMINING ATTRIBUTES OF FACILITIES AND ENTITIES

When it is time to plan for the design of the data bases used in a geographic information system, it is important to identify the attributes of the facilities and entities that are needed in performing the functions of government. Attributes describe the entities of a street (location, size, condition, type of material, date installed, etc.) and are the data elements that are needed in a geographic information system so that they will support a function ("repairing city streets") that will meet one of the goals of the organization ("maintaining the urban infrastructure").

This process is the essence of planning for a GIS: modeling the data so that it accurately represents the facilities and entities that are objects of the functions supporting the long-range goals of the organization. Using the infrastructure example, then, we can define the components of a model that can be used in the long-range plan for a geographic information system in local government (see Table 7.3).

Table 7.3. Components of a long-range GIS plan.

Goal	One of the major strategic directions for the government.
	Examples: Maintain the urban infrastructure
	          Reduce crime
	          Create jobs for the citizens

Function	A major activity within the organization which supports a goal of the organization.
	Examples: Repair city streets
	          Enforce the building codes
	          Assess the value of property

Facility	The physical, legal, or other object upon which a function of government is performed.
	Examples: Street
	          Building
	          Parcel

Entity	Components of the facilities that are managed by the various responsibilities of government and about which attribute data are collected to assist in its management.
	Examples: Pavement
	          Curbs
	          Gutters

Attributes	Descriptive data that define the characteristics of the entity. These attributes answer such questions as where, when, how big, how much, how many, etc.
	Examples: Location
	          Condition
	          Size
	          Date
	          Type

Geographic Information Needs Study

Once the GIS planning phase has identified the goals and functions of local government that utilize geographic information for managing its facilities and entities, an evaluation of existing geographic information sources is needed in order to determine where improvements are needed. This evaluation consists of an inventory of existing maps and geographic information sources and interviews with key personnel who use the maps and information to perform their functions. The interviews are especially helpful for identifying existing problems with the use of geographic information, not only for determining the benefits of a GIS, but also for use later when specific applications are designed.

Since a geographic information system allows different users (offices, departments, agencies, etc.) to share a common base of map information, GIS planners must perform a survey of all government functions to identify the various maps that are used, how they are used, and where they are created and updated. This requires an in-depth study of map utilization on a corporate-wide basis.

A written survey of map users—A map utilization questionnaire completed by the organizational unit responsible for each function of government will record which maps are used, how often they are used, why they are used, what information on them is important, how often they are updated, and how much time is spent in using or updating them.

Oral interviews with key personnel—An interview with the people responsible for using the map in each function is necessary to verify and clarify the information from the survey and to identify existing problems or desired improvements to the maps or their use. Although time-consuming and labor-intensive, these interviews usually uncover surprising (and sometime shocking) problems and inefficiencies in the way maps are used in the organization. One of the most revealing inefficiencies found during these interviews is the process through which many people go in manually copying information from one map to another. (The Map Inventory in Milwaukee identified four different sets of zoning maps—three at different scales in the Planning Department and another in the Building Inspection Department. Whenever a zoning ordinance was approved by the Common Council, each of the four sets had to be updated manually. In a separate interview, it was discovered that the local natural gas company obtained city property maps on a regular basis and redrew them at a different scale to create its own set of maps.)

Review of related documents—Often, a review of existing documents is helpful to GIS planners in understanding why and how maps are used for certain functions. State and local laws usually provide insight into what maps and information are required or mandated from outside the organization. (One map series was found to be mandated by state law when, in fact, it was not needed by any function within the city.) Internal procedures and standards are also useful in understanding why maps are used and updated in certain ways that are not obvious to the uninitiated. In addition, a good source of information about map utilization can be found in the annual budget and its related documents. Usually, quantitative data such as the number of maps processed, the number of maps updated, and the number of hours spent updating the maps are available from these documents.

The map utilization cross-reference chart—While most of the information obtained from these map inventory activities will be retained in the minds and the notes of the GIS planners, formal documentation is required in order for the decision-makers to understand the extent of map usage in the organization. Since top administrators and elected officials do not have the time needed to thoroughly read and study a comprehensive report on the findings of a map inventory, simple presentations must be devised to communicate the information to them. A good method for communicating the extent and amount of map utilization in the organization is a map utilization cross-reference chart, as shown in Figure 7.1.

The map utilization cross-reference chart has a row for each map or use of a map and a column for each function or organizational unit. The information in each cell where a row and column intersect identifies the amount of usage of a map by a function or organizational unit (continual use, frequent use, occasional use, or no use). A chart such as this gives the decision-makers a simple visual understanding of the *extent* of map use in the organization (represented by the number of rows and columns on the chart) as well as the *amount* of map use by the various functions (represented by the density of the symbols in the cells).

Review of existing information systems—One of the most popular misconceptions about geographic information systems is that nongraphics data from the organization's administrative data bases must reside on the GIS with the base map information. The problem with this concept is that many local governments have already implemented information systems on existing computers to support their nongraphics data needs (tax assessment, building permit reviews, crime incident/offense reporting, etc.). Some of these systems may have been operating successfully for many years, and to abandon them in order to implement them on a new GIS would not only disrupt daily operations, but also create additional costs that may not be necessary.

APPLICATION / USER

Application	City Engineer	Planning Department	Building Inspector	Tax Assessor	Traffic Engineering	Election Commission	Cable TV Administration	Health	Mayor's Office	Community Development	Fire	Police	Sanitation	Water Works	Forestry	Common Council	Street and Sewer Maintenance	Municipal Equipment Management	Outside Agencies	Public Information
Quarter Section Mapping	■	■	■	■	■		■	□		◨		◨	◨	◨		◨	◨	■		◨
Construction/Paving Plans	■		■		■												◨			
Curb Lines	■	◨	◨	◨	■					□			■	□					◨	◨
House Number Atlas							■				◨	◨	◨	◨		◨	□			□
Land Use Maps	□	■			◨		◨	□	□		◨	◨	□	□	□	□				
Choropleth Maps	□	■	◨	◨	◨	◨	□	◨	◨	◨	◨	◨	◨	◨	□	□	◨	◨		◨ ◨
Zoning		■	■	■					□								◨		◨	◨
Plan Examination		■	□														□			□
Inspection Workload		■															□			
Violation Mapping	◨	■					◨	◨	□								□			□
Redistricting		■					□	□									□			
Tax Plat Mapping	■	■	■	■			■	□	□		□	□	□	■		■	◨		■	◨
Street Light System		□		■	◨								□			□	□		◨	□
Underground Conduit		□		■	□								□			□	□		◨	□
Traffic Signal Records		□		■	□							□		□		□	□		◨	□
Traffic Control Maps		□		■	□							□		□		□	□		◨	□
Accident Data	◨							□			◨					□			□	□
Election District Maps	◨	◨	◨	◨	◨	■	◨	◨	◨	◨	◨	◨	◨	◨	◨	■	◨	◨	◨	◨
Reapportionment	◨	◨	◨					■								■				□
Cable TV Monitoring	□	◨		◨		■	■		◨					◨		◨	◨	◨	□	
Violation Inspections	◨	■					■		◨							■				□
Inspection Management								■								□			□	□
Violation Mapping		■	■					□			□	□				□			□	□
Arson Investigation								◨	■	■						◨				
Incident Maps	◨							◨	■	■						◨				
Resource Allocation-Fire								◨	■	■						◨				
Crime Statistics								◨	◨	■						◨				□
Resource Allocation-Police								◨		■						◨				
Automated Dispatch										■										
Garbage Collection							◨					■				◨		■	◨	□
Snow Removal								◨		◨	◨	■				◨		■	◨	□
Off Street Parking	◨											■				◨		■	◨	□

Key: ■ = Used Continually   ◨ = Used Frequently   □ = Used

FIGURE 7.1. Map utilization cross-reference chart showing the extent and amount of map utilization among different users.

240

When existing computerized information systems contain nongraphics attribute data that are important for use in a GIS, it is not necessary to abandon them and then recreate them on the new GIS. The data needs interviews with people who use these systems can provide valuable information for determining how a GIS should be implemented and integrated with these existing systems. If many problems are experienced with an existing system, then it may be beneficial to redesign it on the GIS; however, if an existing system provides adequate information service to its users, then an interface between it and the new GIS may be more beneficial by avoiding the expense of redesigning it for the GIS. Interfacing a GIS with an existing system (mainframe, minicomputer, or microcomputer) can be achieved either by transferring files to the GIS or by processing transactions from the GIS directly to the data base of another system and then back again to the GIS.

If appropriate communications hardware and software are installed so that the GIS can be connected to a multisystem network, then direct file transfers (moving entire data bases from an existing system to the GIS) are possible, as are direct data-base accesses between systems. An example of a direct data-base access from a GIS into another computer system is described in the Chapter 3 application, enforcing zoning restrictions. In this application, an address is used by the GIS to find and display its property map. This address is then transferred directly to the property data base of another system to obtain the attribute data for the property. The data are then transferred back to the GIS for display on a tabular report. (See Figure 3.6). If the systems cannot be networked together, then files can still be transferred by periodically copying data onto a tape or disk from one system and then reading the data onto the other system. The boarded-up properties application described in Chapter 3 is an example of this process. In it, a data base of information describing boarded-up properties resides on a microcomputer system and is physically transferred to the GIS (on a diskette), where it is read and then displayed on a map (see Figure 3.26).

There is a number of technologies available to local government for interfacing a new GIS with nongraphics data bases residing on existing computer systems. These existing systems should be studied in depth during the geographic information needs study in order to evaluate their value and determine whether they should be interfaced to the GIS or abandoned and redesigned on the GIS.

Gaining Organizational Support

The easiest way to obtain approval for implementing a geographic information system in local government is to demonstrate that the benefits of using the system will be greater than the costs of implementing it and then supporting it year after year. This can be accomplished by performing a traditional cost/benefit study that identifies and quantifies all of the estimated costs associated with implementing and operating the system and all of the projected dollar benefits that will accrue by using it in future years. If the benefits are large enough to justify the costs, then decision-makers have little difficulty in approving the project.

Often, however, the cost/benefit study is treated exactly for what it is—a study. It offers no proof that the costs will be as estimated and the benefits will be as projected. For this reason, it is valuable to gain some limited experience in using a system before committing the funds to embark on such an ambitious project. This can be done by conducting a pilot project—the limited use of a system during which numerous applications can be tested and cost and benefit projections can be fine-tuned.

A favorable cost/benefit study and successful pilot project may still not be enough to convince the decision-makers that funds should be committed to the project. Some may feel that it will take too long for the benefits to be realized. Others may feel that the project is too ambitious and cannot succeed without additional support. Still others may feel that the technology is too new and are not willing to commit the taxpayers' money to fund, essentially, a research project. When this is the case, GIS advocates must take additional measures to assure the decision-makers that the project can be justified.

The Cost/Benefit Study

The cost/benefit study of a geographic information system is no different from the analysis of costs and benefits of any new computer application in local government—the costs of hardware, software, application programs, and converting the data to digital form must be calculated and compared with the cost savings of other dollar value to the organization once the system is in operation. The GIS cost/benefit study, however, is more difficult than other studies because the initial implementation cost is so high compared with other information systems projects and because the benefits are more difficult to identify since they can be realized in so many different government functions.

COSTS

The high implementation cost of a GIS is caused primarily by the high cost of creating the digital base map of the jurisdiction. With no exception, the map conversion process (or map creation process if no existing maps are available or if existing maps are inadequate for digitization) is much more expensive than its hardware and software costs. Local governments that have completed their digital map conversion process have found that these costs have ranged between 45 and 80 percent of the total cost of the project once it was completed. Depending upon the amount of information required, it can cost between $11 and $27 per land parcel either to digitize from existing property maps or to create digital maps from aerial photography and land surveys in urban areas. (For the City of Milwaukee, these costs were calculated to be $18 per parcel, based upon a total cost of $2,880,000 for digitizing existing tax maps for 160,000 parcels during the years of 1976–81.) In rural areas where the parcel density is low and the type of information needed is different (soils, forestry, hydrology, etc.), unit costs are often represented in dollars per square mile since most of the maps used are polygon-based. Table 7.4, for example, provides a summary of the cost per square mile to prepare digital map files for an 18-square-mile area in the Town of Randall in Kenosha County, Wisconsin. The cost of $8,163 per square mile for the various overlays identified in Table 7.4 were determined as part of a demonstration project coordinated by the Southeastern Wisconsin Regional Planning Commission in 1985 (SEWRPC, 1985).

The hardware and software costs of a GIS, while decreasing over time as new technological improvements by vendors are made, are still

Table 7.4. **Representative base map conversion costs per square mile for a rural area in 1985 dollars.**

System element	Average cost per square mile (1985 dollars)
Relocating and monumenting US Public Land Survey corners, establishing reference marks and related ties, and preparing dossier sheets and attendant certificates	$1400.00
Control survey work, including establishment of state plane coordinates and elevations for monumented US Public Land Survey corners	2,200.00
Preparation of topographic Base Maps and cadastral base sheets (1:2400 scale), including aerial photography costs	1,960.00
Preparation of cadastral maps	1,200.00
Preparation of digital map files:	
Establishment of map projection grid and computer system data structure	62.50
Digitization of water features, flood plain boundaries, and shoreland area boundaries	80.00
Digitization of structure outlines	80.00
Digitization of pavement edges	93.75
Digitization of land ownership parcels	800.00
Digitization of soils inventory	112.50
Creation of land use overlay	150.00
Creation of zoning district overlay	25.00
Subtotal	$1,403.75
Total	$8,163.75

Southeastern Wisconsin Regional Planning Commission (1985), "The Development of an Automated Mapping and Land Information System: A Demonstration Project for the Town of Randall, Kenosha County" (Technical Report Number 30), SEWRPC, Waukesha, WI (Table 7).

**Table 7.5. Cost distribution of implementing a
GIS in local government.**

Cost category	Total cost (percent)
Base Map conversion	46
Management and overhead	16
Hardware and software	15
Support	13
Hardware and software maintenance	6
Supplies	1
Other	3

a significant factor in the total cost of a system. Costs for computers, work stations, plotters, and specialized software packages can range from $500,000 to $2,000,000, depending upon the complexity and planned size of the system. This is usually determined from the long-range plan, which analyzed and prioritized the applications.

Another significant factor in the total cost of a GIS is the cost of the people to implement and support it on an ongoing basis. Since a GIS usually requires the addition of a new computer system to the data-processing inventory of a local government, additional technical support staff must be hired. A description of the staff resources required to support the GIS project is given later. The cost of supporting a GIS can consume between 10 and 15 percent of the cost of the system.

A representative distribution of costs for implementing a geographic information system in local government, based upon a system of 150,000 to 500,000 properties and costing between $2,000,000 and $6,000,000, is given in Table 7.5.

BENEFITS

The benefits associated with using a geographic information system are directly related to the functions identified in the long-range plan because it is these functions that are improved by using the GIS. Generally, these benefits can be grouped into the following categories:

- *Cost reduction*—The decrease in operating expenses of the organization, primarily caused by a savings in time by operating personnel in performing their tasks more efficiently. These savings can be directly related to budget reductions.

- *Cost avoidance*—The prevention of rising costs in the future caused by projected increases in workload. Local governments that are experiencing rapid growth and others that are experiencing a rapid decay of housing and infrastructure have functions that will increase in activity in the future.

- *Increased revenue*—A GIS can increase revenues of a local government by selling data and maps, increasing property tax collections, and improving the quality of data used to apply for state and federal grants. Other increased revenue opportunities are also possible, depending upon the legal, political, and economic environment of the local government.

Cost Reductions. The automated mapping capabilities of a geographic information system contribute most to cost reductions because they improve the productivity of drafters, engineers, and other personnel who perform mapping in local government. These reductions can be large because the daily map updating and facility construction planning functions are labor-intensive and high-volume activities. For example, property maps are updated for a number of different reasons, as described in Chapter 3; all infrastructure construction projects require construction plans for new and replacement pavement, water mains, sewer mains, and other public utilities; special purpose maps and drawings are continually being prepared for site plans, zoning changes, land use plans, voting wards, etc. The automated mapping capabilities of a GIS for these activities have been shown to increase productivity by as much as 25–75 percent over manual methods (see Table 7.6).

The geocoding capabilities of a GIS can also be a large contributor to the reduction of costs in local government. Many of the clerical tasks of local government involve the process of coding address-based records to some geographic area: hospital referrals to public nurses by nursing district, reported crime incidents by police district or census tract, emergency calls for police or fire service to police squad area or fire station, property records to tax assessment district, birth and death records by census tract, etc. By converting the process from a manual method to an automated method using a GIS, personnel time savings can range between 50 and 100 percent.

Redistricting and workload balancing applications can also have a

Table 7.6. Typical mapping and drafting time savings in using a GIS.

Drafting and mapping activity	Time saved (percent)
Updating tax maps	46
Updating zoning maps	32
Updating quarter-section maps	45
Updating official maps	75
Updating sewer maps	50
Preparing background drawings for paving plans	50
Preparing background drawings for sewer and water plans	60

significant impact on reducing the cost of government for two reasons: First, time savings can be realized when changing district boundaries by using the GIS to analyze workload data (such as described in the inspection district example in Chapter 3); and second, it may be possible to reduce the number of service delivery personnel by balancing workload among the remaining personnel. (Whether the cost savings realized by reducing service delivery personnel in this manner can be attributed to the use of a GIS is a debatable topic, as will be described in this chapter.)

Other cost reduction benefits can be associated with GIS technology, depending upon the volume of geographic data-processing functions identified in the long-range plan. For example, a GIS can reduce the costs of preparing official mailings of local government. First, mailing labels for official notices (zoning changes, public improvement notices, special grant and loan program notices, meeting notices, and other special mailings) to residents, property owners, and businesses within a specific geographic area can be produced by a GIS and thus eliminate the dependence on commercial mailing service companies, which charge up to $50 per thousand addresses. (One large city that prepared over 500,000 such mailings per year saved over $25,000 by preparing the mailing labels on its GIS rather than using commercial services.) Additional cost savings on government mailings can be realized by using a GIS to geocode addresses automatically with ZIP + 4 codes. This is because first class mailings with ZIP + 4 codes are reduced by $0.005 per address. For official mailings such as property tax assessment notices, tax bills, special assessment notices, and license renewal notices, the savings by using ZIP + 4 codes can be significant.

Cost Avoidance. Additional benefits can be quantified when it is known that the workload of existing functions will increase in the future or that new functions will be required in the future. By implementing a geographic information system before the additional work is experienced, it may be possible to avoid future expenses that would be needed in order to meet the additional demand. Jurisdictions that expect high levels of physical growth associated with housing and economic development, for example, can project the number of additional personnel that would be required to plan, design, and monitor the construction of the anticipated developments. Additional staff would be required to prepare new property maps, site plans, land use maps, sewer construction plans, water service construction plans, pavement plans, and other geographic records for other public services such as police, fire, and emergency medical responses, solid waste collection, and snow and ice control. By using a GIS for these tasks, however, the number of additional staff required in the future can be minimized. (Time savings such as those shown in Table 7.6 can be used to determine the estimated costs that can be avoided.) Similarly, older jurisdictions not experiencing significant growth may still be facing significant future workload increases because of increasing maintenance and repairs of existing facilities as they age and then require replacement. Depending on the construction materials used for these facilities, pavements may need replacement after 25–50 years; sewer mains after 50–75 years; and water mains after 100–125 years. Long-range capital improvements planning for the replacement of such facilities include estimates for the surveying, mapping, design, and drafting needs associated with their replacement. Thus, projected increases in facility replacement will create a need for additional personnel that can be minimized if a GIS is used for many of the mapping and geographic information needs for implementing the capital improvement plan.

Other cost avoidance benefits can be calculated for more short-term projections. Jurisdictions that are planning to implement computer-aided dispatch (CAD) systems for automatically assigning fire, police, and emergency medical service response units to locations where calls for service originate, for example, will find that a GIS can be used to shorten the time frame for implementing the CAD system because the geographic base file of the GIS can be used to generate the geographic file needed to support CAD functions. In fact, any information system that is planned to be implemented in the future and that requires geocodes in its data base can be implemented more rapidly by using the geocoding features of a GIS. Some local jurisdictions have also avoided future costs by using a GIS to produce more accurate estimates for the amount of earth that must be moved to prepare industrial Land Bank properties for sale. (Land Bank properties are those properties that are owned and improved by local government in order to make them more attractive for new industries to purchase.) Contractors bidding on the work for leveling and grading the land use these figures to estimate their costs, and more accurate estimates of the work create more accurate estimates of the costs.

Increased Revenues. A geographic information system can often be used to increase local government revenues that can be quantified for inclusion in the cost/benefit study. These revenues can be generated from the sale of data to other organizations, charging for geocoding and geoprocessing services, and from collecting additional property taxes that could not be collected without a GIS.

Because a GIS creates new ways to organize, manipulate, and display geographic information, new ways of using public information are created that can be of value to many private companies and other governments. Property-related data and maps can be produced by a local government from its GIS and then sold to private companies for marketing and other activities (see Table 7.7). The spatial analysis capabilities of a GIS can also generate revenues by charging other organizations for geographic information processing services. The simple process of geocoding address-based records to census tract of ZIP + 4 areas can be of great value to organizations that do not have these capabili-

Table 7.7. Potential products and their clients for marketing GIS capabilities to the private sector for increasing revenues of local government.

Product	Potential client

Property ownership lists—Names and addresses of property owners selected and printed by specific geographic area, by type of property, by size of property, or other attribute. These data may also be sold as mailing labels, diskettes, maps, or computer tapes.	Real estate developers
	    Identifying potential properties to purchase
	    Estimating acquisition costs
	Cable and pay TV companies
	    Advertising
	Retail establishments
	    Advertising, especially new businesses
	Contractors
	    Selective advertising to owners of specific types of properties

Property sales—Names, addresses, and other characteristics of recent property sales by geographic area, type of property, etc.	Property appraisers
	    For use in appraising similar properties
	Insurance companies
	    For advertising home-owner insurance plans

Building code violations—Names and addresses of property owners who have been ordered to repair specific deficiencies such as foundation, paint, roof, electrical, plumbing, etc.	Insurance companies
	    To monitor the condition of properties they have insured
	Contractors
	    To identify potential customers needing home improvements
	Financial institutions
	    To identify properties in which they invested and take actions to prevent devaluation

Census and other demographics—Block level statistics and maps of age, income, household composition, ethnic background, and other demographic data	Retail establishments
	    Market analysis for new establishments
	Hospitals and HMOs
	    To determine the need for specific services
	    To identify potential market areas and sites for expansion or relocation

Households or voters—Names and/or addresses of residents in specific areas	Election campaigns
	    For mailing campaign materials and soliciting contributions
	Retail establishments
	    Advertising targeted to specific areas

Digital base maps—Copies of the digital maps can be used by other organizations that have geographic information systems	Utility companies
	    To eliminate the need to create a base map
	Engineering companies
	    To eliminate the need to digitize background information for plans

ties. (One county sheriff paid a large city in the county to use its GIS to group warrants by census tract so that deputies could deliver them more efficiently.) Address verification can also be useful to other governments and credit companies. (One county Welfare Department paid a city to use its GIS to identify the types of properties to which welfare checks were being sent in order to verify that they were being sent to valid residences.) A local government can also receive revenue by charging for providing maps of data from another organization such as dot maps of addresses or choroplethic maps of statistical information. Redistricting capabilities can also generate revenue by using population data to redraw political district boundaries for other governments. (One local school district paid a consultant $8,000 to redistrict its school board representative districts manually utilizing population data. This work could have been accomplished by the city government as a revenue-producing task.) Creative GIS advocates can identify a number of potential markets in their local environments for selling GIS data and services to the private sector and other agencies.

One Midwestern county was able to increase property tax revenues after its GIS was implemented because, for the first time, its property maps could be linked directly to its tax records. After creating its digital property maps and linking their property identifiers to its tax roll, some properties on the map base were found to be missing from the tax roll and, therefore, were not being taxed. Creating a continuous digital map base of properties in the entire jurisdiction allowed this county to collect property taxes on properties that had never been taxed before. (The additional revenues received more than paid for the cost of the GIS!)

While the cost/benefit study contains detailed descriptions and figures associated with the factors contributing to the cost of implementing a geographic information system and the quantifiable dollar benefits gained from using it, there are two effective methods available for summarizing the results of the study for presentation to high-level officials. The first, the *payback period,* is a computation that has been used by private industry for a number of years to evaluate large investment opportunities. The payback period is computed by dividing the total cost of the investment (the cost of implementing the GIS) by the estimated annual return on the investment (the annual benefits of using the GIS). The result is a cost/benefit ratio that represents the number of years it will take to generate enough benefits to pay for the cost of the system. For example, a GIS that costs $1,000,000 to implement and provides $100,000 in annual benefits has a payback period of 10 years (1,000,000 divided by $100,000 per year). This means that, once the GIS is implemented, it will take 10 years to recover the $1,000,000 cost. Basing the decision to invest in GIS technology solely upon its payback period, however, can be ineffective in local government, which has much more pressing needs for funds as local issues and problems are addressed during the annual budget process. Crime, education, welfare, health, and other headline-producing issues (which are not usually analyzed from a cost/benefit point of view) often receive more attention by elected officials who face re-elections in periods that are usually shorter than GIS payback periods.

The second method for summarizing the results of a GIS cost/benefit study is one that emphasizes the annual benefits of using a GIS in graphical form. The *baseline cost comparison chart* as shown in Figure 7.2 is a cost/benefit presentation methodology that this author first saw used by Hank Emery at one of the first annual Keystone Conferences (in the mid-1970s), which are now sponsored by AM/FM International, a professional organization for automated mapping and facilities management systems. The baseline cost comparison chart emphasizes the benefits associated with GIS technology by comparing the costs of running local government *without using* a GIS to the costs of running local government *with* a GIS. The difference between the two is the benefit of a GIS. In Figure 7.2, for example, the solid line represents the costs of mapping and geographic information processing tasks in local government projected over a 10-year time frame using traditional manual methods. The line slopes upward because it represents the cost of tasks that are labor-intensive and therefore always increasing. The dotted line represents the cost of mapping and geographic information processing tasks in local government projected the same 10-year time frame using the projected costs of implementing and using a GIS, with-

FIGURE 7.2. Baseline cost comparison chart for comparing the cost of implementing and operating a GIS to the cost of not implementing and operating a GIS.

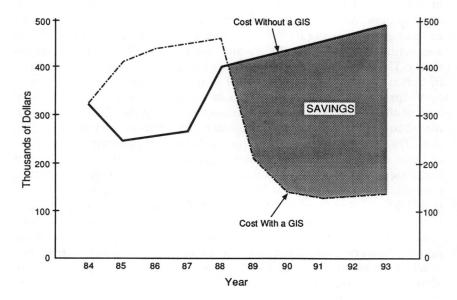

out the labor-intensive costs. When the dotted line is below the solid line, the GIS is producing cost savings over the manual method. When the dotted line is above the solid line, the additional costs for implementing the GIS early in the project are represented. The shaded area, labeled "savings," is a graphical representation of the benefits associated with the use of a GIS. The advantage of representing GIS benefits in this manner is that the baseline of manual costs (the solid line in Figure 7.2) is real, using existing staff or staff that will be required in the future. The dotted line, representing projected benefits of using a GIS, is far enough below the solid line that, even if they are wrong by as much as 50 percent, the savings are still larger than the costs. Representation of costs and benefits of using a GIS by using a baseline cost comparison chart gives elected officials a different perspective on the GIS issue: The costs of running government will escalate more rapidly if nothing is done than if a geographic information system were implemented.

The Pilot Project

A *pilot project* is the limited-term use of a geographic information system using data for a small geographic area to test the planned applications and demonstrate the capabilities of the system to key people in the organization. While most local governments send a small group of officials to other cities and counties to see how existing geographic information systems operate prior to committing time and money to a study, a pilot project that demonstrates GIS applications and capabilities using local maps and information is much more effective in gaining support for the funding of a project. The pilot project will usually answer that final question before the decision is made: "Yes, GIS looks like it works and can improve local government, but will it work *here?*"

The pilot project does not commit the local government to the entire cost of purchasing and implementing a system. It generally commits only about 5–10 percent of the funds needed for the entire system. The best method for doing this is to select a vendor and then award a "lease with option to purchase" contract. This type of contract commits the local government only to the costs associated with leasing the system for a specific time period. Upon successful completion of the pilot project and a final decision to purchase the system, then the system is retained and payment is made. If the results of the pilot project fail to convince decision-makers to fund the project, then the system is returned to the vendor, and no further payment is made. In addition to reducing the obvious financial risk factor, this method allows the project, once it is approved, to proceed directly to the production stage without incurring any time delays associated with acquiring a system and implementing it. It also allows all program and data-base development to be retained without transferring it to a different system. In addition, it allows complete control of the system by the local government staff and provides easy access to it by many people for demonstration purposes. Unfortunately, the "lease with option to purchase" clause is becoming increasingly more difficult to write into contracts because GIS vendors are not as willing to accept them as they were in years past when the technology was still unproved.

If it is not possible to obtain a contract with such a clause in it, there are other methods for conducting a pilot project without committing the organization to the full cost of a system. It may be possible, for example, to purchase or lease one or two work stations and connect them to an operating GIS in the area. This would require arrangements to be made with another government, public utility, or engineering company that has a system to lease time on their system. If that is not possible, then a pilot project could be conducted "off-site" at a company that provides GIS services for a fee.

Whatever method is used to obtain the use of a system for a limited period of time, the pilot project must be systematically planned and conducted so that the time available is used most productively and efficiently. The pilot project must have a plan with clearly defined objectives, a schedule of tasks and assigned responsibilities, and measurable criteria for evaluating the results of the test.

Objectives. While the goal of the pilot project is to prove the value of GIS technology in a specific organization, there are six major objectives that should be set in order to achieve that goal:

Test the feasibility of each application. The GIS applications identified in the long-range plan should be tested during the pilot project to determine if they work as envisioned. After trying each application, its feasibility can be determined by reviewing the costs and benefits of implementing and using it, the ease of using the system for the application, and whether it is a suitable use of GIS technology.

Refine the cost and benefit estimates. There is no better way to estimate costs and benefits than by actually performing the tasks that will eventually generate those costs and benefits. Converting maps and data into digital form for a small geographic area will generate much more realistic cost estimates than using figures obtained from other jurisdictions that have different needs and different sources of data. This will allow "real life" examples to be conducted so that accurate figures can be computed, such as the time saved in creating a drawing, updating a map, or retrieving information.

Demonstrate the capabilities of the system. Key decision-makers do not have the time to travel to another city or county to see a GIS in operation. The pilot project makes it easy to schedule many different people for demonstrations and also allows them to observe the applications and capabilities of the system on maps and data familiar to them.

Determine user acceptance. Many people are apprehensive about using new technology to perform tasks that have been familiar to them for many years. Since the success of the system depends so much on the people who will be using it on a daily basis, it is of critical importance that they become convinced that it will help them perform their tasks better. Thus, problems with user acceptance that may be encountered during the pilot project can be addressed and corrected immediately before they adversely affect the project once it is approved. Also, some planned applications may be dropped from the long-range plan at this stage if the people performing them do not feel comfortable using the system.

Select the best methods for implementation. In addition to testing the planned applications, the pilot project also allows the organization to test different methods for making them work. One of the first decisions that must be made concerning the implementation of a system is how to create the base map. The pilot project allows a low-risk environment to try different methods: digitizing existing maps, contracting with commercial services to digitize existing maps, using aerial photography and coordinate geometry, or even combinations of different methods. The pilot project period also allows programmers to test different programs, user commands, and menus to perform certain tasks. Tests on different data structures and definitions of layers can also be valuable during the pilot project.

Spark user imaginations. A valuable source of ideas for using GIS technology is the group of people who will be using the system once it is implemented. Since these are the people who best know the details of the tasks that must be performed in local government functions, they are also the ones who will best know how GIS technology can be used to improve those tasks. Once they see GIS capabilities in action processing the information they work with every day, they will suggest new applications that were not obvious during the planning phases of the project.

Tasks, Assignments, and Schedules. As with most limited-term projects, the pilot project will yield the best results if each task is clearly defined, scheduled for completion at a specific date, and assigned as a responsibility to a person or office. (There are many microcomputer-based project management software packages available from the private sector that can aid in the management and administration of a pilot project to make these activities easier.) Table 7.8 is an example of a plan that was used successfully by a large city to conduct its pilot project.

Evaluation. Since the pilot project is a period of testing new technology and new procedures, it is best to establish tasks within the pilot project plan that will evaluate the status of these tests at certain points during the pilot period. This will allow adjustments to be made during the trial period that can improve further testing or identify new tests to be made before the trial period is over. For a 6-month pilot project, a typical evaluation interval is every 2 months.

Table 7.8. Example of a pilot project plan for a GIS.

Task 1.0 Receive delivery of the system, install, test, and accept hardware and software.

Task 1.1 Receive training for programmers and digitizers, including command menu use, detailed digitizing, and operational support of the system.

Task 2.0 Create city map index to establish a continuous coordinate system and to establish a record-keeping system for tracking the progress of the map conversion process.

Task 3.0 Create base map for one quarter-section. The first of 13 quarter-section maps will be digitized in accordance with the detailed digitizing document. To ensure that the procedures are sufficient to maintain good quality control, two digitizers will digitize the same map concurrently. A hard copy plot will be produced at this point and will be compared with the original tracing to ensure that the map was converted as precisely as possible. Upon completion of the quarter-section map conversion, we will add the additional information from the official map that does not duplicate the information previously digitized from the quarter-section map. Next, information from the corresponding tax plat maps will be added.

When this first quarter-section base map has been digitized and meets the acceptance criteria, then an evaluation will be made to determine the time estimates required to complete one quarter-section. Based upon these estimates, it will be determined if it is feasible to complete the remaining 12 quarter-sections in the pilot project area. If necessary, we will reduce the number of quarter-sections to be completed in the pilot.

Task 3.1 Once the first quarter-section base map is completed, the procedures should be well tested and the digitizers should be familiar enough with using the system so that the remaining quarter-sections can be converted more rapidly, simulating a map conversion production environment that will help refine the time and cost estimates of the map conversion process. This task can be accomplished in parallel with many of the tasks to be described.

Task 3.2. When a quarter-section is completely digitized and checked for accuracy, it will enter into a maintenance cycle so that any changes that occur to the original tracing will be changed on the data base as well. This will provide data that can be used to compare updating time and cost of manual versus computer methods.

Task 3.3. Background drawings for public works construction projects will be created from the digital base map for sewer, water, and pavement construction projects in the pilot project area.

Task 4.0 The property file interface will be tested by extracting the records of all properties in the pilot project area from the tax assessor's data base and loading them onto the data-base management system of the GIS. These records will be matched on tax key number with the digital base map to establish the coordinate linkages between the maps and the data.

Task 4.1 Specific spatial analysis functions will be tested from this property file interface: retrieval of a map and attribute data for a specific address; identifying a land parcel on the screen and receiving a display of the property attributes for the selected parcel; displaying a symbol at each property having a certain attribute value; calculating statistics on property attributes (counts, totals, averages, etc.) for all properties within an arbitrarily determined polygon; and shading predetermined polygons to create choroplethic displays.

Task 5.0 Sewer information will be added to the digital base map for the pilot project area. Two map sources will be used: the Sewer System map and the House Connection Atlas. To accomplish this, it will be necessary to develop standard symbols and line symbologies to represent the various sewer sizes, types, manholes, taps, and other appurtenances.

Task 5.1 Water information will be added to the digital base map for the pilot project area. The Water Distribution Maps will be digitized and symbols and line symbologies eatablished in order to represent the various hardware and main sizes for the Water Atlas. The attribute information, such as the number of turns and length of wrench for the valves, will be established in the data base for use by spatial analysis functions.

Task 5.2 Zoning map information will be added to the digital base map for the pilot project area. The three separate zoning maps (use, area, and height) will be digitized on different layers and then combined onto one map to evaluate the polygon overlay capabilities of the GIS.

The information needed to evaluate the GIS during the pilot project consists of:

1. The time required to digitize the base map;
2. The time required to add each additional map from existing sources;
3. The difference between the manual method for updating the map information and the automated method;
4. The difference between the manual method for retrieving or creating maps and drawings and the automated method;
5. The costs of operating the system over an extended period of time;
6. Savings in time of using the basic drafting features of the system;
7. Evaluation of the nonquantifiable benefits of a GIS.

Justifying the GIS Project

When the results of a cost/benefit study show that it would be economically beneficial to implement GIS technology and when a pilot project demonstrates that the use of a GIS can be valuable to a local government, it still may be difficult to obtain approval for the funds needed to implement the project. This situation can frustrate the GIS advocates who not only believe in the technology, but also feel that their analyses have been conducted in a professionally objective and comprehensive manner. They wonder: "If other organizations are successfully using a GIS and our numbers show a favorable cost/benefit ratio or payback period here, why can't we get funds?" When this happens it is helpful to look at factors other than the numbers and reports supporting the project:

- The cost/benefit study may show that the projected benefits are significantly greater than the estimated costs, but the numbers used are often those of the GIS advocates, who may not be the managers of the people who will be performing the work and therefore, responsible for creating the benefits. When the people who perform the cost/benefit study are not the same people who must generate the benefits, conflicts can arise that may cause uncertainty in the eyes of the decision-makers.

- Those who *do* have the responsibility to generate the potential benefits have a different perspective on the issue. For one, their mission is to complete a task and not necessarily to use the system. If the system cannot help them in achieving their mission, they will not use it. Also, managers are often reluctant to reduce the number of employees under their direction because there always seems to be more work to assign to those whose tasks have been improved. It is possible, therefore, that a GIS can generate savings in the time it takes to perform a task, but if there is other work to do, budget reductions may not be realized.

- Those who will be responsible for updating the data base may feel they are giving up control over the information. Since the power of a geographic information system is its ability to centralize data from many different functions and integrate it geographically for use by many other functions, control of data is dispersed among many people, rather than held by the few who used to control separate data bases or manual files. This situation can make some key project participants feel threatened, even when appropriate security controls are incorporated into the design of the system. (One Tax Commissioner, for example, disabled all access to property records on a GIS when she discovered that hundreds of government employees were overloading the system by comparing the assessed value of their properties with those of comparable properties in their neighborhood. Also, a Building Commissioner refused to participate in a data integration effort because he feared that data errors would be found and used to criticize the management of his department. After he retired, however, his replacement turned 180 degrees on the issue by allowing data access to anyone—even the public. He welcomed assistance in finding data errors so that they could be corrected.)

- Those who make the final decision on funding a GIS project may use factors other than the numbers in the cost/benefit study to make their decision. They may have a use for the system that, in the perspective of the long-range plan, may seem insignificant. They may hear about the value of a GIS from colleagues in other organizations, professional publications, or directly from vendors themselves.

- The benefit estimates will be challenged by the Budget Office. This is because the savings that can be generated by using a GIS may not be directly related to reductions in the annual budget. Take, for example, an application that can be used by four different offices and will generate 25 percent savings of time in each office. This does not result in the elimination of one position because each person has other responsibilities. Another example is taken from the workload balancing application described in Chapter 3. While the use of a GIS allowed the workload of building inspectors to be more evenly distributed and, in fact, showed that one fewer inspector was needed, the same analyses could have been performed manually to reach the same conclusion. Should the GIS get credit for producing the savings associated with eliminating an inspector, or can it only get credit for coming to that conclusion more quickly?

These paragraphs describe the intangible factors that can affect the decision to fund a GIS project. Since geographic information systems technology is relatively new and not well understood by local government officials, gaining their support for funding a project is a matter of selling the product. To do this, GIS advocates must create a market for GIS services by promoting their value in performing the functions of local government. Promoting the value of a GIS is much the same as promoting the value of any new information system in local government today because many systems have already been implemented that automate operational functions, and their value has generally been realized through productivity improvements rather than usefulness for information access and manipulation. Martin (1976) addresses the problem of promoting the usefulness of new systems and offers 12 recommendations for systems designers to ensure their value:*

> It will be increasingly difficult in future systems to find tangible justification. For that matter the telephone systems of a corporation, and many other services, cannot be justified tangibly. The justification of a computer doing payroll depends on the number of clerks it replaces. The justification of an information system depends on the *value of the information it provides*. It becomes necessary to assess this value, rather than merely to assess cost reductions or displacements. The value of the information must exceed the cost of providing it.
>
> Can we measure the value of information?
>
> The answer to that question depends on what the information is used for. If it is used in an "operations system," the value of the information can usually be estimated, at least approximately. If it is a general-purpose information system, the value may be impossible to estimate in any other than the most subjective terms. It would be exceedingly difficult to estimate the value of having your telephone. Information systems will be evaluated equally subjectively.
>
> To have the maximum likelihood of being valuable to its potential users, computer-provided *information* must have a number of characteristics, which are listed [below].
>
> 1. It must be *accurate*. The worst criticism of some information systems is that the information is inaccurate.
> 2. It must be *tailored to the needs of the user*.
> 3. It must be *relevant* to what he requires at that time.

*James Martin, *Principles of Data-Base Management Methodologies*, © 1976, pp. 292–294. Reprinted by permission of Prentice Hall, Inc., Englewood Cliffs, New Jersey.

4. It must be *timely*. Often it must be given in response to a user's request. If it is given to him a day late, he may not use it in some cases.
5. It must be *immediately understandable*. Some computer printouts are remarkably unintelligible.
6. Its significance must be immediately *recognizable*. This is often a function of the method or format or presentation.
7. It helps if it is *attractively presented*.
8. It should be *brief*. The lengthy listing characteristic of batch processing often conceal rather than reveal information. Single significant facts should not be camouflaged by the inclusion of other less relevant data.
9. It should be sufficiently *up-to-date* for the purpose for which it will be employed.
10. It should be *trustworthy*. Management is often suspicious of computerized information sources. Management will soon lose confidence in them if occasional errors in the data are found.
11. It should be *complete*. The user should not be left feeling that he has received only part of the information he really needs. . . .
12. It should be *easily accessible*. If a terminal is difficult to use, or confusing, it will not be used.

Managing the GIS Project

After funds are finally approved to implement the geographic information system, the project takes a different focus. Instead of studying the potential applications and promoting the value of GIS technology throughout the organization, GIS advocates find themselves responsible for the successful implementation and management of the system. This means that they must begin to create the digital base map that will be used for the many different applications to be implemented, and they must assemble the appropriate staff and organizational support structures to ensure the successful operation of the system on a long-term basis.

Base Map Creation

There are a number of different choices available to local government for creating the digital base map of its geographic information system. Not only do these choices involve the *source* of the information but also the *method* for converting the information to digital form. The Chapter 2 discussion on automated mapping, for example, described how *existing maps* can be either digitized or scanned to create the digital base map. The discussion on cadastral maps and planimetric maps in Chapter 6 explained how *new maps* can be created from either aerial photography or from the legal descriptions recorded in nongraphical form. Chapter 5 described the contents of the DIME and TIGER Files, which are *digital maps* available from the U.S. Census Bureau for direct input to a geographic information system. Depending upon the quality of the source and the applications planned in the long-range plan, the GIS base map can be created from any one or a combination of these different sources and methods. Usually, the pilot project is used to determine the most appropriate approach for any given jurisdiction.

Existing Maps

Since creating new maps is a major undertaking that can cost millions of dollars (even with the use of GIS technology), local officials are wise to first evaluate their existing maps to determine if they are adequate to use as the source of the digital base map for a GIS. This evaluation should consider the following factors:

1. Are they accurate? Converting existing maps to digital form will not make them more accurate—it will only amplify the inaccuracies and no one will use them.
2. Are they current? If months or even years of updates and changes have been piling up in some corner because of higher-priority work, it may make better sense to create new maps than to try to catch up with the changes.
3. Are they stable? It is unwise to convert maps that have been stored for many years on paper or some other unstable medium because their size may have changed as a result of humidity, temperature, or other environmental influence, causing distortions in map features.
4. Are they clean? Maps with folds, stains, dirt, and other marks are difficult to convert to digital form because it is hard to distinguish between map features and unwanted markings.
5. Are they tied to a continuous coordinate system that is based upon a geodetic reference? Maps that have been used for many years as individual map sheets and have not been tied into a coordinate system or geodetic control network will not fit together when they are converted to a continuous digital base map of the jurisdiction.

If existing maps can meet these tests, then it may be better to use them in the map conversion process to avoid the added expense of creating new maps for the digital base map.

As described in Chapter 2, there are two methods for converting existing maps to digital form: digitizing and scanning. *Digitizing* is the time-consuming process of redrawing the map by using an electronic cursor with the map affixed to a digitizing table (see Chapter 2). It is the most common method currently used to convert property maps because scanning technology has not yet improved the efficiency of adding dimension and other textual data that are abundant on such maps. *Scanning* existing maps through a raster-to-vector optical scanner is much faster than digitizing but requires an editing process afterwards to ensure that the information was converted correctly and to eliminate any superfluous marks that may have been converted. Maps with little text and dimension data require less editing and therefore are more amenable to scanning than dense property maps.

NEW MAPS

Local governments that do not have maps or whose existing maps do not pass the tests listed above may use either aerial photographs or existing legal survey records to create the digital base map for their GIS. As described in Chapter 6, aerial photographs are used to create planimetric maps that identify the locations of physical features that can be seen on the photographs. Legal survey records are land survey- ors' notes, legal descriptions, and coordinate values that can be used in coordinate geometry (COGO) software to create a digital map. In many cases, both methods must be used in order to store both physical and cadastral features in a base map.

Creating a base map from aerial photographs is called *photogramme- try*. It is an expensive process because it involves a number of steps to create the digital map:

1. Ground control—Monuments to be identified on the photograph must first be surveyed to record their coordinates, which are used later in the process.
2. Marking features—Physical features that are small (less than 1 or 2 feet) must be marked with paint or some large material forming a larger feature so that they can be seen easily on the photograph.
3. Flying and photographing—The actual process of taking photo- graphs from airplanes requires careful planning to consider fac- tors such as tree foliage, snow cover, shadows, altitude, and other situations that can affect the ability to identify features on photo- graphs.
4. Stereo compilation—After the photographs have been developed and printed, they must be converted to orthophotographs in or- der to eliminate the distortions caused by the curvature of the earth and the angle of view.

The resulting orthophotographs can then be used either to digitize the observed features (including those that have been marked by paint or other material) to create the digital base map, or they can be scanned directly to create a digital orthophotograph in raster format. This image can be used as a background for display and for vectorizing desired features as part of an editing process in building the digital base map. Storing the rasterized image of an orthophotograph can be very expen- sive, however, since each pixel on the photograph must be assigned a coordinate pair.

The cadastral information that cannot be observed in a photograph can then be added to the base map by using coordinate geometry software from the nongraphics records of legal descriptions and surveyors notes. This process is usually difficult (as described in Chapter 6) because the legal records must use the same monuments for survey control as are identified in the photographs (usually, these monuments are the Public Land Survey System monuments).

DIGITAL MAPS

Other governments, agencies, and even private companies can be sources for base maps already in digital form. The TIGER File described in Chapter 5, for example, is available from the U.S. Census Bureau for a nominal fee. Another digital base map, the Digital Line Graph (DLG), is available from the U.S. Geological Survey as part of their National Digital Cartographic Data Base (Southard, 1984). Neither the TIGER Files nor the DLG Files contain parcel map data, however, and therefore can be used only in small-scale GIS applications such as block or census tract statistical and geographic analysis. Their low price (usually for the cost of reproduction) and widespread transferability to most commercially available geographic information systems makes them quite popular in many local governments.

Digital base maps containing parcel boundaries are sometimes shared among different local governments and agencies such as utility companies. While some GIS projects include a consortium of different organizations that share the cost and also contribute data to construct a digital base map, other jurisdictions have been able to successfully transfer a parcel-oriented digital base map from one system to another. The Wisconsin Electric Power Company, for example, obtained a digital base map from the City of Milwaukee to avoid converting their own maps in the Milwaukee portion of their service area. The cost of purchasing the digital information from the city was much lower than the estimated cost to digitize their own maps.

Private companies also provide digital base maps to local governments on a commercial basis. These files are usually byproducts of the primary service they provide: producing maps, providing GIS software and services, or converting maps to digital form for other organizations. This is because so many private companies are already using GIS technology in their normal operations. Ironically, some local governments that have not yet implemented GIS technology may already have a digital base map available to them because of the activities of other organizations in the area.

Organizing People to Implement and Operate the GIS Successfully

With a potential of millions of dollars to be invested in a multiuser geographic information system, the most important resources needed to ensure that the funds are appropriately used are the people responsible for implementing and operating the system. Hardware, software, and data-base problems can be identified and addressed (albeit with additional time and money); however, staff who are lacking in the necessary technical and communicative skills, motivation and dedication, and creativeness and willingness to accept new concepts, will most likely ensure failure because neither time nor money can be enough to overcome these deficiencies.

Brown (1989) has identified ten general roles that people with these characteristics must perform. Realizing that few governments can afford ten staff positions for these roles, Brown acknowledges that, often, one individual takes on more than one role and, "in many cases, numerous people perform the same one."

Manager. The GIS Manager is responsible for managing both the GIS project team and the system itself after it is implemented. This person understands how the GIS can improve the organization and communicates with the managers in user departments to ensure that the GIS can meet their needs. During the study stages of the project, the GIS Manager must take a marketing and educational role in explaining the benefits of GIS technology to nontechnical people. Later, during the pilot project, base map conversion, and implementation stages, the GIS Manager must focus on coordinating resources and managing the project workplan.

Analyst. This person utilizes specific technical knowledge and experience in applying GIS technology to solve particular user problems and satisfy their information needs. The Analyst must interview potential users, design specific programs and procedures, ensure that they are implemented correctly, and train the users in their operation. The Analyst must not only possess GIS technical skills, but also be a good communicator who can listen to users and translate their needs into GIS programs and procedures that can be used to generate products from the system.

Systems Administrator. Once the system is installed and operating correctly, the GIS project team must ensure that the hardware and software continues to function correctly on a daily basis. The Systems Administrator is responsible for maintaining the system in a continuous operational mode, responding to and solving problems as they occur. The Systems Administrator monitors system performance, installs new versions of software and additional hardware and software components, performs periodic backups (saving data on external media), maintains adequate security on the system, and ensures that a sufficient amount of supplies are available as they are needed. With the enhancement of GIS technology towards local area networks (LANs), in which many computers are linked together to share data files and other resources, the GIS Administrator must also understand LAN communications technology and manage the many files stored on the system.

Programmer. The GIS Programmer translates the application specifications prepared by the GIS Analyst into programs, user menus, and macro-level commands to perform specific functions needed by the users of the GIS. Together with the GIS Analyst and the users, the GIS Programmer also assists in developing new data bases needed for additional applications.

Processor. The GIS Processor is a "Super User"—one who knows all of the capabilities of the GIS hardware and software and can utilize them to produce specific products needed by the users. This position is especially important to the organization that cannot afford to provide GIS work stations to all functions needing GIS products and services. Departments and functions that do not have their own work stations can use the services of the GIS Processor when they need specific products on an ad hoc basis. This position is valuable to a GIS project team because the GIS Analyst and Programmer can become too far removed from the detailed needs of the users and therefore must rely on the GIS Processor to produce specific maps and reports from the system.

Data-Base Administrator. Most data-processing organizations have a person who performs data-base administration functions to assure that the logical design of data bases is appropriate for the requirements of the physical storage of data and the data-base management software. The GIS Data Base Administrator assists the Analysts, Programmers, and Users organize geographic features into layers, identify sources of data, develop coding structures for nongraphics data, and document information about the contents of the data bases so that others may know what information is available on the system.

Cartographer. Users, as well as GIS project teams that provide services and products to users, may require someone who is skilled in cartography to design map displays that are clear, are understandable, and deliver the intended message. The Cartographer also assists in the development of standard map symbols (much like the Data-Base Administrator, who develops standard data coding structures) and establishes standard map series for general distribution.

Drafter. Most user departments provide GIS training to people who are experienced in drafting to take advantage of their knowledge of mapping and drafting principles. The Drafter is especially needed during the pilot project and base map conversion stages of a project to compile and integrate cartographic data from many sources in preparation for the digital conversion process.

Digitizer. The labor-intensive process of converting map information into digital form requires a number of people who can work for long periods of time at the digitizing table or at the work station screen. Digitizers must pay careful attention to details when they digitize or edit map information. While drafting skills are beneficial to the Digitizer, it is much more important for the Digitizer to have significant familiarity with the mapping and drafting standards and conventions of the particular local government.

End Users. Those who take advantage of the capabilities of GIS technology to meet the geographic data-processing needs of the functions of local government are called End Users (or Users, for short). They are the ones who make the system produce the benefits that were originally anticipated during the planning and evaluation stages of the GIS project. It is for this reason that it is important to educate and train End Users on the capabilities of the system—including new capabilities as they are implemented. It is also for this reason that it is important to involve the End Users early in the development of the project and keep them informed as the project progresses. This will prevent unrealistic expectations as well as minimize the fear that is often associated with the introduction of new technology in the workplace.

Local governments that have been most successful in implementing and using GIS technology have been able to assemble staff members with the skills necessary to assume these roles. In addition, the success of these efforts has been shown to be dependent upon two other common characteristics: strong GIS leadership and continuity of staff.

GIS leadership, as with most other forms of leadership, involves a dedication to an achievable goal and the ability to motivate others towards achieving that goal. Since GIS technology can have a profound effect on the way a local government manages and uses its geographic information (see the quote at the beginning of this chapter), it is important to have a strong leader who can help the many managers, policy-makers, and operations personnel understand and adjust to a new tool for improving their jobs. A GIS leader will devote a considerable amount of time and energy explaining GIS capabilities to key personnel at the top as well as the bottom of the organization chart. He or she has a keen ability to listen and understand how and why work is accomplished in the organization, with a unique ability to see relationships among functions and data in the organization—without placing GIS technology at a more important level than the work itself. The GIS leader can also communicate, interpersonally as well as formally, with people at all levels of the organization without intimidating the nontechnical people with GIS or data-processing jargon and without communicating in a condescending manner. Internally to the project team, the GIS leader motivates staff members by clearly defining his or her vision and sharing successes (as well as failures) at specific stages of the project. The GIS leader can come from outside the project team (such as a department head or elected official), or the GIS leader may well be the GIS Manager of the team.

By maintaining the continuity of staff members of the GIS project team, both technical expertise and knowledge of the internal processes and personalities of the organization are retained on a long-term basis, which improves stability and productivity of staff members. If key people (especially the GIS Manager) leave the project and are replaced by new people over the development of the project, the priorities of the project can change and delays can occur. Not only does it take time for new people to learn the particular processes and procedures of a local government, but the interpersonal relationships between potential users and the GIS project team must be re-established. Brown (1989) emphasizes the need for staff continuity: "Successful sites keep these key individuals. This appears to be fundamental to their success. When one or two of these key people depart, they take with them your system's experience, knowledge, and inertia." (p.3)

While leadership and staff continuity must persist throughout the multiyear development of a GIS project, specific skills for all ten roles described earlier are not needed at the onset of the project. As the project proceeds through its various stages (study, implementation, and operation), different skills are needed at different times.

THE STUDY PHASE

Analysis, education, and marketing skills are most important during the evaluation of needs and project justification stages of a GIS project. The GIS Manager, Analyst, End Users, and even consulting services are needed during this stage to conduct the long-range plan, information needs study, map inventory, and cost/benefit study. While it is not important to establish a formal office or organization to house the project team at this stage, it is important for the team members to be assigned full time to the project so that other distractions will not interfere with progress. This means that Budget personnel assigned to the project should remain on their project tasks even during the months in which the budget is set; data-processing personnel assigned to the project should remain on their project tasks even during tax bill preparation and year-end accounting processing; and representatives of policy-level personnel assigned to the project should remain on their tasks even during election campaigns.

Since a GIS can affect many different functions in a local government and since a formal organization is not usually established for what could be only a short-term project, it is important to establish a Steering Committee of high-level department managers to review project progress and establish priorities during the study phases of the project. The GIS Steering Committee, established early in this stage and continuing through the pilot project and base map conversion phases, can assure an integrated design of the system as well as universal acceptance of the system by many different departments. The GIS manager should submit a project work plan and report progress against that plan to the Steering Committee on a regular basis to keep members informed and to identify important policy issues early so that they may be resolved promptly.

THE IMPLEMENTATION PHASE

Analysis, programming, drafting, digitizing, and GIS processing skills are needed once the pilot project and base map conversion phases begin. Since this stage involves rather specific tasks with critical target dates and deliverable products, the GIS project team must become more formalized into an official, centralized office within the organizational structure of the local government. Exactly where the GIS project should be placed organizationally within the government is a major policy decision that the Steering Committee must make at this stage. On the one hand, it is desirable to place control of the project in the organization that has the most responsibility for data update and maintenance, while, on the other hand, control of the project should be placed in an office that has a corporate-wide perspective on the organization. Some local governments divide the responsibilities for the project between those offices that are responsible for the data (Public Works, Tax, Planning, etc.) and the office responsible for the hardware and software (Data Processing, Administration, etc.). Other local governments that rely on consultants for the pilot project and base map conversion utilize a centralized office to oversee the necessary contracts and act as a liaison between user departments and the contract services. However the responsibilities are assigned (centralized, shared, or contracted), the GIS Steering Committee should remain a strong influence, meeting regularly during this stage to review progress and expenditures and to resolve policy issues.

THE OPERATIONAL PHASE

Management, data-base administration, cartography, programming, and systems administration skills are most important after the system has been implemented and the base map conversion process completed. Once the GIS has developed into a mature, operational service to the local government, it is necessary to establish appropriate management and administrative mechanisms for long-term operations. At this stage of development, the GIS project team takes on many of the characteristics of a data-processing office, with responsibilities for operational support, applications development, data-base and network management, and information retrieval services. In addition, the GIS office becomes the centralized point of focus for implementing applications in accordance with the long-range plan. Since the political environment of the government and local issues change over time, the GIS long-range plan must be reviewed and updated on a continual basis—preferably as part of the budget process.

The primary focus of the project at this stage is maintaining communication among the various users of the system and providing necessary security and data standards control over the centralized data base. The Steering Committee is no longer needed at this stage because the management of the GIS is incorporated into the organizational structure of the government. It is normally replaced by a User's Group, consisting of End Users from the many different functions of the government that use the system. The User's Group provides a channel for communicating changes to the system (hardware and software), new applications and capabilities added, problems with using the system and administrating the data base, and any other information that needs to be disseminated among users and between users and the central GIS support staff. The GIS User's Group should meet at least on a monthly basis, managed by the users themselves, and attended and staffed by the central GIS support staff.

Another important issue that must be addressed during the operational stage of a system is cost recovery, or how to share the costs of the central GIS support staff (which can approach the hundreds of thousands of dollars per year). Should the cost of the GIS support staff be shared among the users, or should the cost be included in the overall administrative costs of the government (much the same as Personnel, Legal, Purchasing, and other central support services)? This is an issue that is really part of the general issue of data-processing cost recovery in local government. In its simplest form, operational cost sharing can be prorated by dividing the total cost of the centralized costs (hardware and software maintenance, support staff, and supplies and other overhead costs) by the number of work stations connected to the system to determine a cost per work station each user pays. In a more complex form, operational cost sharing can be based upon a formula that considers the activity of use (number of data-base accesses) and the size of the data bases stored for access.

Table 7.9 summarizes the resources required at the different stages of GIS development in local government.

References

Brown, Clint (1989), "Implementing a GIS: Common Elements of Successful Sites," paper presented at the 1989 Annual Conference of the Urban and Regional Information Systems Association, Boston, Mass. (pp. 1–3).

GIS WORLD (May, 1989), Vol. 2, No. 3, Fort Collins, Colo. (p. 16).

Martin, James (1976), Principles of Data-Base Management Methodologies, Prentice-Hall, Englewood Cliffs, N.J. (p. 293).

SEWRPC (1985), "The Development of an Automated Mapping and Land Information System: A Demonstration Project for the Town of Randall, Kenosha County," Technical Report Number 30, Southeastern Wisconsin Regional Planning Commission, Waukesha, Wisc. (p. 59).

Southard, R.B. (1984), "Automation of the National Cartographic Data Base: Implications to State and Local Governments," Seminar on the Multipurpose Cadastre: Modernizing Land Information Systems in North America, Bernard J. Niemann, Jr. (ed.), Wisconsin Land Information Reports: Number 1, Institute for Environmental Studies, University of Wisconsin—Madison, Wisc. (p. 266).

Table 7.9. GIS resources required for each stage of GIS project development. Full-time responsibilities depend upon the complexity of the system.

Stages of project development	Leader	Manager	Analyst	GIS processor	Data-base administrator	Programmer	System administrator	Cartographer	Drafter	Digitizer	End user	Contractor/consultant	User group	Steering committee
Study														
Long-range plan	X		X								X	X		X
Information needs study	X		X								X	X		X
Map inventory	X		X								X	X		X
Cost/benefit study	X		X								X	X		X
Implementation														
Pilot project	X	X	X	X	X	X			X	X	X	X		X
Base map conversion	X	X	X	X	X			X	X	X	X	X		X
Applications development		X	X	X	X	X		X	X		X	X		X
Operation														
Data-base management		X	X	X	X	X	X	X	X				X	
Network management		X	X			X	X						X	
Operations support		X		X	X	X	X		X	X			X	
Cost recovery		X	X				X						X	

GIS Resources

Typical staff size by complexity of system

Large multi–user system
Centralized minicomputer
Small microcomputer network
Single microcomputer

ADDITIONAL READINGS

Bullen, Christine V., and Rockart, John F. (1981), *A Primer on Critical Success Factors*, Center for Information Systems Research, Working Paper No. 69, Sloan School of Management, MIT, Cambridge, Mass.

Croswell, Peter L. (1988), "Definition of Applications as a Basis for GIS Planning and System Procurement," papers from the annual conference of the Urban and Regional Information Systems Association, 1988 (Volume IV), URISA, Washington, D.C.

Huxhold, William E. (1980), "Planning Agencies and the Data Processing Department: Working Together Effectively," *Computers in Local Government: Urban and Regional Planning*, AUERBACH Publishers, Inc., Pennsauken, N.J.

Joffe, Bruce A. (1988), "A Strategy for Implementing Automated Intelligent Mapping," papers from the annual conference of the Urban and Regional Information Systems Association 1988 (Volume II), URISA, Washington, D.C.

Martin, James, with Leben, Joe (1989), *Strategic Information Planning Methodologies* (Second Edition), Prentice-Hall, Englewood Cliffs, N.J.

Applications Digest

The following pages include a list of functions of local government and associated tables of applications of geographic information systems in city and county governments across the United States. The tables are by no means exhaustive, but are a result of months of research by Kathy Bertrand, a graduate student at the University of Wisconsin—Milwaukee, who researched many journals, publications, papers, and information provided by vendors of geographic information systems.

While no standard taxonomy of GIS applications is generally accepted in the industry at the present time (an effort at the National Center for Geographic Information and Analysis, however, is currently underway), these applications have been organized by general categories of generic local government functions that are listed below.

Where information was not available on a specific user or application, a dash appears in the appropriate column. In many cases, the available information was based upon the recorded plans of an organization; however, actual implementation was not verified. Accordingly, these applications have been listed here and identified as "(future)."

Functions of Local Government

Health and Safety	City/County	User	Applications
Building codes	Mesa (AZ)	Fire Department	Building inspection
	Alhambra (CA)	Building and Safety	Permit information
			Licensing
			Inspection analysis
			Code enforcement
			Redevelopment
	Fresno County (CA)	—	Building codes
	Orange County (CA)	—	Code enforcement
	Redlands (CA)	—	Permit processing
	San Bernardino (CA)	—	Permit processing
	Chattanooga (TN)	Public Works Department	Inspections
	Alexandria (VA)	Housing and Codes Enforcement Agencies	—
	Arlington (VA)	Inspection Management	Permits
		—	Inspection/code enforcement
			Housing condition zoning
	Roanoke (VA)		
	Bellevue (WA)	Public Works	Building permits
			Permit tracking
	Green Bay (WI)	Municipal	Permits/inspection/revaluation
	Milwaukee (WI)	Planning Department	Thematic mapping of housing quality
		Building Inspection Department	Permit processing
			Identifying building code violators
			Inspection workload balancing
			Housing code violations
Disaster preparedness	Los Angeles (CA)	City Planning Department	Emergency preparedness planning
			Dam inundation
			Shelter allocating/critical facilities
			Hazardous structures recovery

Health and Safety	City/County	User	Applications
			Hazardous materials mapping (future)
			Damage incidence reporting (future)
		City Administration	Disaster preparedness
			Risk management
	San Bernardino (CA)		Flood control
		Property Information Management	Hazardous areas
	Springfield (IL)	Emergency Services and Disaster Agency	—
	Frankfort (KY)	Disaster and Emergency Services	—
	Jefferson City (MO)	Emergency Management Agency	—
	Brooklyn/Queens/ New York City, (NY)	Transit Authority	Subway evacuation/rescue
	Dutchess County (NY)		Evacuation routes
			Hazardous material spills
	Tonawanda (NY)	Police Department	Hazardous chemical spills
	Pittsburgh (PA)	Planning Department	Risk assessment
	Garland (TX)		Transportation of hazardous materials
	Toronto (Canada)	Fire Department	Fire incident analysis
	Mesa (AZ)	Fire Department	Fire incident analysis
			Street maintenance
			Emergency response analysis
Fire services			—
	Phoenix (AZ)	Fire Department	Fire Department inspections
	Alhambra (CA)	Fire Department	General operations
	Anaheim (CA)	Fire Department	Specific analysis

Location	Agency	Applications
Beverly Hills (CA)	Fire Department	General planning
		Hazardous materials
Burbank (CA)	Fire Department	Fire analysis
Long Beach (CA)	Fire Department	Hazardous materials
Los Angeles (CA)	Fire Department (future)	Inspections
Orange County (CA)	Fire Department	Hydrant location optimization
San Bernardino (CA)	Forestry/Fire Warden	Utility emergencies
	—	Research/prevention/safety
		—
San Diego (CA)	Fire Department	Computer-aided dispatching/emergency response system
Denver County (CO)	Fire Department	Response zones (future)
Schaumburg (IL)	Fire Department	Inspection areas (future)
Minneapolis (MN)	Fire Department	Firefighters/dispatchers
Washoe County (NV)	City of Reno Fire Department	Response time analysis
Newark (NJ)	Fire Department	Incident analysis
		—
Broome County (NY)	Fire Department	Locations of hazardous waste materials
Freeport (NY)	Fire Department	Incident tracking
Manhattan (NY)	Fire Department	Medical records/related data
Suffolk County (NY)	Fire Department	Fire incident analysis
		—
		—
Dayton (OH)	Fire Department (future)	Districts
Washington, DC	Fire Department	Hydrant locations
Chattanooga (TN)	Fire Department	Dangerous substances location
		Emergency equipment location
		—
		Emergency vehicle dispatch
		—

Health and Safety	City/County	User	Applications
	Austin (TX)	Fire Department	Call volume prediction for annexed areas
			Resource allocation
			Accident response analysis
	Alexandria (VA)	Fire Department	Emergency activities
	Roanoke (VA)	Fire Department	—
	Bellevue (WA)	—	Hydrant locations
			Emergency (future)
	Milwaukee (WI)	Fire Department	Emergency vehicle response areas
Health services	San Bernardino (CA)	Public Health Department	Patient distribution analysis
			Animal control dispatch
	Detroit (MI)	Medical Agency	Planning medical facilities
	Broome County (NY)	Health/Hospitals	Medical facilities planning
	Suffolk County (NY)	Health Department	Disease outbreak patterns
			Planning facilities
			Water quality monitoring
	Milwaukee (WI)	Health Department	Public Health nursing districts
			Health incidents
Police services	Vancouver (Canada)	Police Department	—
	Victoria (Canada)	Saanich Police Department	—
	Montreal (Canada)	Urban Community Police Department	—
	Mesa (AZ)	Police Department	Social indicators (future)
			Crime analysis (future)
			Dispatching (future)
	Phoenix (AZ)	Maricopa County Sheriff's Office	—
	Alameda County (CA)	Police Department	—

Alhambra (CA)	Public Safety	Crime patterns analysis
Anaheim (CA)	Police Department	General operations
		Specific analysis
		General planning
		Dispatch
Beverly Hills (CA)	Police Department	Calls/crime analysis
Burbank (CA)	Police Department	Strategic planning
		Incident analysis
		Strategic planning
Carson (CA)	City Sheriff's Station	—
Fullerton (CA)	Police Department	—
Garden Grove (CA)	Police Department	—
Hawthorne (CA)	Police Department	—
Huntington Beach (CA)	Police Department	—
City of Industry (CA)	Sheriff Station	Crime analysis
Long Beach (CA)	Public Safety—Police	Inspection/control
Los Angeles (CA)	Police Department	Tactical planning
		Technical services
Martinez (CA)	Countra Costa County Sheriff	—
Monterey Park (CA)	Police Department	—
Norwalk (CA)	Sheriff's Department	—
Pasadena (CA)	Police Department	—
San Bernardino (CA)	Sheriff's Department	Law enforcement analysis
		Computer-aided dispatch
		Crime analysis
San Clemente (CA)	Police Department	—
San Diego (CA)	County Sheriff's Department	—
San Leandro (CA)	Police Department	—
	Alameda Sheriff's Office	—

Health and Safety	City/County	User	Applications
	Stockton (CA)	Police Department	—
	Union City (CA)	Police Department	—
	West Covina (CA)	Police Department	—
	Boulder (CO)	Police Department	—
	Denver County (CO)	Department of Public Safety	Arrest statistics
			Criminal defense process
			Vehicle accidents/drunk driving data
	Osceola County (FL)	Osceola County Sheriff	—
	West Palm Beach (FL)	Police Department	—
	Cobb County (GA)	Cobb County Police Department	—
	Schaumburg (IL)	Police Department	Response time analysis
	Indianapolis (IN)	Police Department	—
	Minneapolis (MN)	—	Crime analysis
	St. Paul (MN)	—	Police call locations
	Jackson (MS)	Sheriff's Office	Central dispatch
	St. Louis (MO)	Police Department	—
	Washoe County (NV)	City of Reno Police	—
	Newark (NJ)	Police Department	Crime reporting/crime analysis
		Municipal Courts	Criminal/transportation analysis
	Broome County (NY)	Police Department	Crime analysis
	Freeport (NY)	Police Department	—
	New York City (NY)	Police Department	—
	Suffolk County (NY)	Police Department	Criminal activity analysis
	Syracuse (NY)	Police Department	Crime analysis
			Resource allocation
	Tonawanda (NY)	Police Department	Dispatching emergency vehicles (911)
			Handicapped–special requirements

Safety services		
Dayton (OH)	Police Department (future)	Car theft locations
Oklahoma City (OK)	Police Department (future)	—
Salem (OR)	Police Department	Police dispatch
Washington DC	Police Department	Emergency vehicle dispatch
Chattanooga (TN)	Metro Police	—
Austin (TX)	Police Department	Call volume prediction for annexed areas Resource allocation Accident response time analysis
Alexandria (VA)	Police Department	Emergency activities
Arlington (VA)	Police Department	Crime incident analysis
Chesapeake (VA)	Police Department	—
Bellevue (WA)	Police Department	Criminal records (future)
Milwaukee (WI)	Police Department	Property maps for beat patrol Crime statistics mapping/analysis (future)
Conway (AK)	Office of Emergency Services	—
Alhambra (CA)	Public Safety	Hazardous waste material location
Burbank (CA)	—	Dog licensing/shelter locations (future) Hazardous materials (future)
Huntington Beach (CA)	—	Public safety
Los Angeles (CA)	City Planning Department Department of Animal Regulation	Location of handicapped persons Licensing/control/care
Los Angeles County (CA)	Emergency Operations	—
San Bernardino (CA)	Communication	911 system
San Diego (CA)	—	Public safety (future)
Colorado Springs (CO)	—	Emergency services
Hernando County (FL)	—	Geocoding/incident reporting
Schaumburg (IL)	—	Emergency services

Health and Safety	City/County	User	Applications
	Indianapolis (IN)	Department of Public Safety	—
		Department of Civil Defense	—
	Sedgwick County (KS)	Emergency Communications	—
	Silver Spring (MD)	—	Emergency operator response/inquiry
	Jackson (MS)	—	Emergency management
	Dutchess County (NY)	—	Pollution emittance
	Dayton (OH)	—	Emergency vehicle dispatch
	Chattanooga (TN)	—	911/emergency management (future)
	Nashville (TN)	Emergency Management Agency	—
	Austin (TX)	—	Emergency services
	Alexandria (VA)	—	Facilities management
	Bellevue (WA)	—	Emergency services
	Green Bay (WI)	—	Emergency services
Sanitation	Burbank (CA)	—	Garbage routing/maintenance (future)
	Long Beach (CA)	Public Works	Refuse routing
	Fort Lauderdale (FL)	—	Solid waste collection
	Minneapolis (MN)	City Council	Solid waste siting
	St. Paul (MN)	—	Hazardous waste disposal siting
	Newark (NJ)	Administration	Sanitation/landscape planning
	Broome County (NY)	—	Garbage collection
	Chattanooga (TN)	Public Works Department	Garbage collection routing/scheduling
	Garland (TX)	Department of City Development	Landfill site selection
	Milwaukee (WI)	Public Works Department	Solid waste collection-route balancing

Public Works	City/County	Users	Applications
Buildings and structures	Toronto (Canada)	Property/Management Services Dept.	Design/renovate buildings
	Compton (CA)	—	Building permits
	Huntington Beach (CA)	—	Building permits
	Los Angeles (CA)	Department of Building & Safety	Building permits
			Investigation/conservation
	San Diego (CA)	—	Building/development permits
	Schaumburg (IL)	—	Building location site analysis
	Garfield Township (MI)	—	Building permits
	Chattanooga (TN)	Public Works Department	Building plans
			Permits
			Proposed construction
Forestry services	Green Bay (WI)	—	Tree trimming
	Milwaukee (WI)	Public Works Department	Forestry (future)
Transportation	Mesa (AZ)	Department of Transportation	Street maintenance/inventory
			Vehicle scheduling
			Vehicle routing
	Alhambra (CA)	Police Department	Traffic incident analysis (future)
	Burbank (CA)	Public Safety	Traffic accident analysis
		Department of Transportation	Parking facilities
			Traffic counts
			Route analysis
	Huntington Beach (CA)	—	Road maintenance
			Traffic engineering
	Long Beach (CA)	Public Works Department	Street routing
	Los Angeles (CA)	Department of Transportation	Traffic zone analysis
			Planning/operations
			Transportation planning

Public Works	City/County	Users	Applications
	San Bernardino (CA)	Police Department	Traffic coordination
		Bureau of Engineering	Street inventory
		Department of Transportation	Vehicle routing
			Traffic analysis
		Airports	Traffic management
	San Diego (CA)	Department of Transportation	Circulation plans (future)
			Inventories (future)
			Street-related permits (future)
			Traffic data (future)
			Route data (future)
			Dispatch records (future)
			Street networks (future)
			Street related facilities (future)
			Road characteristics (future)
	Colorado Springs (CO)	—	Transportation
	Orange County (FL)	—	Routing functions
			Street inventory
	Ada County (ID)	Ada County Highway District	Car pooling (future)
		—	Road maintenance analysis (future)
	Du Page County (IL)	—	Traffic information
	Schaumburg (IL)	Transportation/Parking	Traffic characteristics
			Traffic demand
			Inventory
			Traffic analysis
			Parking requirements
			Road improvements
			Pavement design
	Indianapolis (IN)	Department of Transportation	—

City/County	Department	Applications
Billings (MT)	Department of Transportation	Creation of maps while you drive
Washoe County (NV)	—	Traffic zone analysis
Newark (NJ)	Administration	Demolitions/streets & paving
New Brunswick (NJ)	Department of Transportation	Transportation planning
		Trip generation
		Trip production/attraction
Broome County (NY)	—	Bus route planning
New York City (NY)	Metro Transit Authority	Transportation planning
		Highway and transit assignments
		Subway analysis
		Improvements/forecasting ridership
Suffolk County (NY)	Public Works Department	Traffic safety
		Bus route planning
Coldwater Village (OH)	—	Traffic
Hamilton County (OH)	County Engineer's Office	Accident statistical analysis
Oklahoma City (OK)	—	Transportation/traffic planning
Columbia (SC)	—	Transportation
Chattanooga (TN)	—	Transportation (future)
		Street paving/maintenance analysis
Austin (TX)	—	Transportation
Dallas (TX)	Dallas Area Rapid Transit Authority	Completely automated system
		Regional trends/ridership analysis
		Rail planning/design
		Facilities management
		Thematic mapping
		Bus planning
		Scheduling/operations
		Rail planning
		Customer information/bus schedules
		Car pooling service

	City/County	Users	Applications
Public Works	Garland (TX)	Department of Transportation	Infrastructure improvements (future)
	Roanoke (VA)	Department of Transportation	Traffic analysis
	Bellevue (WA)	Department of Transportation	Traffic accident indicators
			Access roadways
			Accident records
			Street inventory
			Traffic distribution
	Madison (WI)	Department of Transportation	Highway/road/street networks
	Milwaukee (WI)	Bureau of Traffic Engineering (future)	—
Utilities	Kenai (AK)	Public Works Department	Water/sewer/utilities management
	Mesa (AZ)	Public Works Department	Utilities
	Phoenix (AZ)	Water and Waste Water	Water and waste water management
		Salt River Project	Electric distribution facilities management
	Tempe (AZ)	Public Works Department	Water/sewer management planning/ coordination
	Anaheim (CA)	Utility Department	General operations/analysis/planning
	Burbank (CA)	—	Infrastructure management
		Administration	Meter route optimization
		Public Service–Electric/Water Division	Electronic transformer load analysis
			Municipal/utilities inventory
			Flow/demand analysis—water
		Engineering Department	Sewer flow projects/modeling
	Huntington Beach (CA)	Municipal/Utility Companies	—
			Infrastructure maintenance
	Long Beach (CA)	Public Works Department	Sewer/street maintenance
			Facilities management

Los Angeles (CA)	Community Development Department	Water/power distribution/planning
	Bureau of Engineering	Facilities management/planning (future)
		Utility management
San Bernardino (CA)	Airports	Facilities management
	Department of Transportation	Facilities management
	Engineering Department	Sewer/septic facilities
San Diego (CA)	Utility Department	Water utilities ownership
		Utility development planning
		Utility characteristics (future)
		Telecommunications (future)
Colorado Springs (CO)	Public Works/Planning	Utility construction/operation/ maintenance/distribution (future)
Commerce City (CO)	Adams County Communication Center	—
Denver County (CO)	Public Service Company	New material management (future)
		Gas distribution facilities (future)
		Real estate (future)
Clearwater (FL)	—	Utility network locality/monitoring
Brevard County (FL)	—	Utilities
Gainesville (FL)	—	Facilities management
		Service delivery
		Utility planning
		Infrastructure management/inventory
Orange County (FL)	Public Works Department	Utility distribution
		Water distribution
	—	
St. Petersburg (FL)	Public Works Department	Hydrolic performance of utility networks
		Utility system analysis
Winter Park (FL)	East Central Florida Regional Planning	Utility planning

Public Works	City/County	Users	Applications
	Boise (ID)	—	Utility ownership (future) Utility maintenance (future)
	Chicago (IL)	—	Wastewater treatment/water control facilities Utility/facility locations
		Department of Maintenance and Operations	Monitor/manage/operations of facilities
	Schaumburg (IL)	Storm/sanitary sewer/water distribution	Inventory Line location Facilities analysis
	Frankfort (KY)	—	Public/private utilities Facilities management
	New Orleans (LA)	Parish Communication District	—
	Silver Spring (MD)	—	Infrastructure management Maintenance management Facility planning decisions
	Boston (MA)	Massachusetts Water Resources Authority Boston Harbor	Maintenance and capital improvement management Public water supplies
	City of Hopkins (MN)	—	Utility monitoring/location
	St. Paul (MN)	—	Highway/powerline routing
	Jackson (MS)	Mississippi Public Service Commission	Facilities/infrastructure management
	St. Louis (MO)	—	Utility network modeling applications Facilities management projects Infrastructure management
	Billings (MT)	—	Utility status in new subdivisions Utility/street inspections

Location	Department	Application(s)
Clark County (NV)	Sanitation/Water Department	Facilities Pole locations
Washoe County (NV)	Utility Department	Infrastructure management
Concord (NH)	—	Facility siting (future)
Newark (NJ)	—	Water supply/maintenance needs
Freeport (NY)	Public Services	Utility management/monitoring/maintenance Electric distribution Street lighting
New York City (NY)	Department of Environmental Protection	Manage/rebuild water/sewer system
Suffolk County (NY)	Public Works	Line locations Utilities
Dayton (OH)	Sewer Maintenance (future) Water Department	— —
Cincinnati (OH)	Metropolitan Sewer District	Automated mapping Location of sewers/drainage studies Storm water management utility Maintaining/upgrading storm sewer system System modeling/break history Facility inventory/location Restore/maintain infrastructure Pressure studies
	City Water Works	Utilities Phone cables Stormwater/sanitary sewer Water demand/monitoring
Coldwater Village (OH)	—	Tracking/managing/mapping/analyzing land-based information/facilities
Hamilton County (OH)	—	Infrastructure construction/maintenance

Public Works	City/County	Users	Applications
	Philadelphia (PA)	Metropolitan Sewerage District	Automated mapping
		—	Waste/stormwater collection
			Facility/maintenance/inventory/main break information
	Columbia (SC)	—	Automation of utility operations
			Water supply systems
			Wastewater systems
			Infrastructure improvement
	Lexington County (SC)	—	Automation of utility operations
	Richland County (SC)	—	Automation of utility operations
	Chattanooga (TN)	Public Works Department	Stormwater facility needs
			Utility property easement assessment analysis
			Urban & storm drainage retention analysis
			Street closure/utility repair location
			Infrastructure management/planning/ maintenance
			Utility networks (future)
			Composite utility map generation (future)
	Austin (TX)	—	Utilities
			Facilities location analysis
			Development monitoring
	Dallas (TX)	Municipalities/Public Utility Companies	Utility location/monitoring/maintenance
	Denton County (TX)	Electric Cooperative	—
	Alexandria (VA)	—	Facilities management

City	Department	Uses
Arlington (VA)	Department of Public Works	Street/sewer/water locations Infrastructure management
Portsmouth (VA)	Public Works Department	Water system upgrade sewer (future) Rehabilitating infrastructure (future)
Roanoke (VA)	—	Sewer/water
Bellevue (WA)	Utility Department	Utilities Facility inventory Utility meter routes Maintenance planning Sewerage planning
Seattle (WA)	King County Communications	—
Green Bay (WI)	Municipal Utility Department	Infrastructure management Network analysis Work order generation Inspection scheduling System planning/design Facility distribution/load management Substation information/diagrams Land holdings Easements/base map information Electric transmissions facilities Gas load/flow analysis Gas/electric storm/sewer restoration Load/voltage studies Cathodic protection Duct occupancy Area load forecasting Meter reading routing
Madison (WI)	—	Utility locations
Milwaukee (WI)	—	Water/sewer network mapping (future)

Public Works	City/County	Users	Applications
Planning	Kenai (AK)	Public Works	Planning
	Burbank (CA)	—	Graphic analysis–land-use/network modeling
	Fresno County (CA)	Planning/Public Works	Project surveillance
	Gainesville (FL)	Public Works	Land use
	Clark County (NV)	Public Works	Land net
	Cincinnati (OH)	Public Works Department	Area analysis
	Arlington (VA)	Department of Public Works	Work distribution
	Bellevue (WA)	Public Works	Problem analysis
			Planning/projections
Engineering	Huntsville (AL)	Engineering Department	Surveying
			Cartographic mapping
	Pheonix (AZ)	Engineering Department	Map updating
	Tempe (AZ)	Engineering Department	Public information
	Alhambra (CA)	Public Services	Planning/engineering
	Anaheim (CA)	Public Works Engineering	General operations
			General planning
	Long Beach (CA)	Engineering Department	Road design
	Los Angeles (CA)	Bureau of Engineering	Real estate
			Street construction
			Wastewater permits
	Colorado Springs (CO)	Engineering Department	—
	Orange County (FL)	—	Right-of-way
	St. Petersburg (FL)	Engineering Department	Utility design/maintenance
	Dallas (TX)	Engineering Department	Street lighting/paving
			Maintenance of networks
			Street inventory/maintenance
			Thematic mapping

Minneapolis (MN)	Engineering Planning and Design	—
St. Louis (MO)	—	Civil engineering project alternatives
Newark (NJ)	Engineering Department	Inventory/sewerage/contracts
Riverhead (NY)	Engineering Department	—
Dayton (OH)	Engineering Department (future)	—
Cincinnati (OH)	Engineering Department	—
Hamilton County (OH)	County Engineer's Office	Street/road inventory Priority infrastructure location Infrastructure improvement Identification of site restrictions
Chattanooga (TN)	Engineer's office	Analysis/design Mapping/information retrieval Land attribute data
Austin (TX)	—	Engineering
Utah County (UT)	Engineering Offices (future)	—
Green Bay (WI)	—	Mapping
Milwaukee (WI)	Engineering Department	Map updating
Rhinelander (WI)	Engineering Department	Facilities inventory
Toronto (Canada)	Department of Public Works	—
Alhambra (CA)	Public Services	Physical feature inventory
Anaheim (CA)	Public Works/Engineering	Specific analysis
Hamilton County (OH)	Department of Public Works	—
Chattanooga (TN)	Public Works Department	Municipal information Information retrieval/land attribute data Proposed construction Demographic analysis
General — Utah County (UT)	Public Works Department (future)	—
Bellevue (WA)	Public Works Department	Zoning

Recreation and Culture	City/County	Users	Applications
Cable Television Services	Indianapolis (IN)	—	Utility/Cable TV
	Milwaukee (WI)	Office of Telecommunications	Cable TV (future)
Community events	Milwaukee (WI)	Department of City Development	Special event planning
Historic/Cultural Preservation	San Bernardino (CA)	Property Information Management	Scenic routes
Recreational Facilities	Alhambra (CA)	Recreation	Facility location inventory
			Area market analysis
	Burbank (CA)	Park/Recreation	Site analysis
	Long Beach (CA)	Park and Recreation	Control/reporting capabilities
			Site analysis
	Suffolk County (NY)	Park/Recreation	Facilities
	Roanoke (VA)	—	Park service area
Education	Tucson (AZ)	Tucson Unified School District	—
	Elk Grove (CA)	Elk Grove Unified School District	—
	Lodi (CA)	Lodi Unified School District	—
	Los Angeles (CA)	Los Angeles Unified School District	—
	Orange County (CA)	Orange Unified School District	—
	Pleasanton (CA)	Pleasanton Schools	—
	Pomona (CA)	Pomona Unified School District	—
	Sacramento (CA)	Grant Joint Union High School District	—
	San Diego (CA)	San Diego Unified School District	—

Location	Organization	Application
San Juan Capistrano (CA)	Capistrano Unified School District	—
Colorado Springs (CO)	Colorado Springs Public Schools	—
Lakewood (CO)	Jefferson County Public Schools	—
St. Lucie County (FL)	School Board	Bus routing
Anne Arundel (MD)	Anne Arundel County Public Schools	School redistricting
Ellicott City (MD)	Howard County Public Schools	—
Cambridge (MA)	Education	Pupil transportation routing
Jackson (MS)	Jackson Public Schools	School location analysis
Great Falls (MO)	Great Falls Public Schools	—
Broome County (NY)	Education	—
Wake County (NC)	Public Schools	School enrollments
Canyon (TX)	Canyon Independent School District	—
Denton County (TX)	Denton Public Schools	—
Galena Park (TX)	Galena Park Independent School District	—
San Antonio (TX)	North East Independent School District / San Antonio Independent School District	—
Roanoke (VA)		School water service
Virginia Beach (VA)	Virginia Beach City Public Schools	—
Bellevue (WA)	School System	Bus routing (future)

Urban Development	City/County	Users	Applications
Business regulation	San Bernardino (CA)	—	Licensing
	San Diego (CA)	—	Business license
	Orange County (CA)	—	Building permit distribution
			Permits by census tract
	Orange County (FL)	—	Licensing
	Milwaukee (WI)	Common Council	Liquor licensing analysis
			Special area mailing (future)
Economic development	Phoenix (AZ)	Development Services	Information services
	Alhambra (CA)	Redevelopment Agency	New business impact
	Burbank (CA)	—	Feasibility studies
			Cost/benefit analyses
			Strategic plans
	Redlands (CA)	—	Fiscal impact
	Schaumburg (IL)	Transportation/Parking	Site analysis
			Development analysis
	City of Louisville (KY)	Corporate real estate planners	Corporate real estate opportunities
	Boston (MA)	Real estate developers	Real estate development/property inventory
			Real estate supply/demand analysis
	Cincinnati (OH)	—	Vacant lot inventories
	Cleveland (OH)	—	Vacant lot inventories
	Chattanooga (TN)	Public Works Department	Economic development (future)
			Regional economic development (future)
	Milwaukee (WI)	Department of City Development	Economic development
Environmental amenities	Huntsville (AL)	—	Natural resource impact assessment
	Mesa (AZ)	Planning and Development Department	Resource allocation
	Fresno County (CA)	—	Environmental management
	Huntington Beach (CA)	—	Resource management

Location	Department/Office	Applications
San Bernardino (CA)	Environmental Health	Environmental analysis Resource management Monitoring Inventory Hazardous waste
San Diego (CA)	Environmental	Land use/land cover (future) Flood plains/seismic lines assessment (future) Natural resources (future)
Colorado Springs (CO)	Environmental	Environmental assessment
Denver County (CO)	Environmental Protection Agency Office	Identification of sources of particular pollutants Enforcement of environmental regulations Flood plain analysis
Orange County (FL)	—	Environmental resource analysis
Winter Park (FL)	—	Mine siting
Champaign (IL)	—	Soil analysis Ecology
Schaumburg (IL)	—	Environmental information Resource planning decisions
Silver Spring (MD)	—	Public decision-making Resource protection
Amherst (MA)	Department of Landscape Architecture Regional Planning	
East Lansing (MI)	—	Soils Precipitation/temperature Agricultural analysis Flood plain program
City of Hopkins (MN)	—	Thematic mapping
St. Paul (MN)	Department of Natural Resources	Soil impact Mining impact Erosion assessment
Billings (MO)	—	Crop evaluation

Category	City/County	Users	Applications
Urban Development	Washoe County (NV)	—	Natural features Hydrography
	Concord (NH)	Natural Resource/Regional Planning	River corridor management/watershed (future) Resource identification (future)
	Dutchess County (NY)	—	Environmental management Wetland/stream corridors/soils/aquifer Ground water information
	Cincinnati (OH)	—	Flood plain analysis
	Oklahoma City (OK)	—	Groundwater research Water quality management
	Pittsburgh (PA)	—	Environmental assessment
	Chattanooga (TN)	—	Environmental land-use planning (future)
	Norris (TN)	Tennessee Valley Authority	Reservoir planning Assessment of environmental impacts Water resources
	Austin (TX)	—	Environmental assessment
	Roanoke (VA)	—	Soil/flood plain zoning
Harbor Management	Boston (MA)	Public Works Department	Third Harbor Tunnel Project Alignment of new highway/tunnel structures Risk assessment/contaminant concentration
		Boston Harbor	Boston Harbor protection
Housing	Burbank (CA)	Planning/Redevelopment	Housing
	Los Angeles (CA)	Housing Authority	Real estate Property management
	San Bernardino (CA)	Housing Department/Community Development	Permits
	Colorado Springs (CO)	—	Housing

Category	Location	Agency/Department	Function(s)
	Dayton (OH)	Housing Inspector	Housing conditions; Housing inspections
	Austin (TX)	—	Housing
Land-use control/mapping	Huntsville (AL)		Thematic assessment mapping
	Kenai (AK)	Kenai Peninsula Borough	Manage land development
	Mesa (AZ)	Planning and Development Department	Land-use analysis; Property inventory
	Phoenix (AZ)	Salt River Project	Land ownership
	Little Rock (AK)	Planning and Development Department	Permit tracking/monitoring; Rezoning; Annexations
	Burbank (CA)	Planning/Redevelopment	Compilation of land areas
	Fresno County (CA)	Planning Commission	Zoning; Parcel map ordinances; Subdivision codes
	Long Beach (CA)	Planning and Development Department	Zoning changes; High-density development
	Los Angeles (CA)	City Planning Department	Land-use zoning (city-wide); Land-use/planning
	San Bernardino (CA)	Community Development Department	Land-use planning
	San Diego (CA)	Office of Surveyor	Land use
	Aspen and Pitkin (CO)	—	Land-use analysis
	Wilmington (DE)	—	Vacant lot inventories; Real estate research
	Brevard County (FL)	—	Zoning
	Clearwater (FL)	—	Regional topographical and cadastral mapping
	Orange County (FL)	—	Thematic/statistical mapping; Zoning; Land development; Cadastral parcel/geodetics

Urban Development	City/County	Users	Applications
	St. Petersburg (FL)	—	Comprehensive land-use planning
	West Palm Beach (FL)	Regional Planning Council	Updated land use/land cover by district
	Winter Park (FL)	—	Existing land use
	Schaumburg (IL)	Land-Use/Environment	Land use/zoning
			Site analysis
			Comprehensive planning
			Permits
			Special districts
			Development controls/regulatory devices
	Indianapolis (IN)	—	Land base conversions
	Overland Park (KS)	Planning Commission Department	Land development permits
	Frankfort (KY)	—	Land-use planning
	Boston (MA)	—	Vacant lot inventories
	Town of Hadley (MA)	—	Land preservation
			Zoning
			Landscape inventory
			Land use/land cover
	East Lansing (MI)	—	Zoning
			Elevation
	City of Hopkins (MN)	—	Zoning/land use
	Minneapolis (MN)	—	Auditing zoning maps
	St. Paul (MN)	—	Site analysis
			Existing land use
	Jackson (MS)	—	Public land survey information
	Clark County (NV)	Planning and Development Department	Zoning status
	Washoe County (NV)	—	Land parcels
			Zoning
	Suffolk County (NY)	Department of City Development	Zoning

Category	City	Department	Applications
	Dayton (OH)	—	Real estate activity/zoning; Parcel/property ownership
	Hamilton County (OH)	County Auditor	Plat changes
	Austin (TX)	—	Zoning changes; City limit/annexation boundaries; Public notification of zoning changes (future)
	Alexandria (VA)	—	Real estate research; Vacant lot inventories
	Arlington (VA)	Department of City Development	Zoning; Land-use planning/management
	Roanoke (VA)	—	Land-use analysis
	Bellevue (WA)	Planning	Land-use/zoning
	Green Bay (WI)	—	Vacant and agricultural land tracking
	Milwaukee (WI)	Department of City Development	Land-use mapping; Vacant lot inventories
Neighborhood preservation	Los Angeles (CA)	Department of City Development	Neighborhood/redevelopment
	Redlands (CA)	—	Community analysis
	Schaumburg (IL)	—	Neighborhood information
	Milwaukee (WI)	Department of City Development	Neighborhood improvement
Planning	Toronto (Canada)	Department of City Development	Collecting/maintaining zoning information; Planning subdivisions
	Huntsville (AL)	Department of City Development	Site information; Existing land use; Dwelling description; Land valuation; Location analysis
	Mesa (AZ)	Department of City Development	Licensing; Permits; Physical planning; Zoning assessment

Urban Development	City/County	Users	Applications
	Phoenix (AZ)	Department of City Development	Development services Strategic planning
	Alhambra (CA)	Department of City Development	Parcel information
	Anaheim (CA)	Department of City Development	General operations Specific analysis General planning
	Burbank (CA)	Planning/Redevelopment	Long-range forecasting Land-use trends/analysis
	Fresno County (CA)	Planning Commission	Policy/land development Land development
	Huntington Beach (CA)	—	Planning
	Los Angeles (CA)	Department of City Development	Community Plan Revision Program Specific plans
	Redlands (CA)	Department of City Development	Site profiles
	San Diego (CA)	Department of City Development	Tax rate (future) Zoning (future) Site profiles
	Hernando County (FL)	Planning and Zoning Department	Evaluate property sales analysis
	Orange County (FL)	—	Development planning Comprehensive planning Subdivision approval
	Champaign (IL)	—	Planning Site analysis
	Du Page County (IL)	Department of City Development	Zoning Land-use planning
	City of Louisville (KY)	Department of City Development Jefferson County Metro Sewer District	Site planning Waterfront development project

	Administrative Units	Planning areas
Hennepin County (MN)	Department of City Development	Market property value analysis
Minneapolis (MN)	Department of City Development	Building value analysis Parcel character identification Area characteristics/analysis
Washoe County (NV)	Natural Resource/Regional Planning	Statewide comprehensive planning (future) Planning/growth management (future)
Concord (NH)	Municipal	Comprehensive planning (future)
Dutchess County (NY)	—	Planning
Freeport (NY)	—	Planning
Riverhead (NY)	Department of City Development	—
Suffolk County (NY)	Department of City Development	Zoning
Cincinnati (OH)	City Planning Department/Land Use	Subdivision processing
Dayton (OH)	Department of City Development (future)	—
Hamilton County (OH)	County Regional Planning Commission	Development planning
Oklahoma City (OK)	—	Comprehensive planning
Columbia (SC)	—	Industrial site selection
Chattanooga (TN)	—	Industrial development site selection (future)
Austin (TX)	Capital Area Planning Council	City comprehensive planning
Garland (TX)	Department of City Development	Long-range forecasting
Utah County (UT)	Department of City Development (future)	—
Arlington (VA)	Department of City Development	Site plan system
Portsmouth (VA)	—	Inspection management Capital improvements planning (future)
Roanoke (VA)	—	Comprehensive planning

303

Urban Development	City/County	Users	Applications
	Green Bay (WI)	Department of City Development	Zoning Development analysis City planning
	Milwaukee (WI)	Department of City Development	Planning/urban renewal Community development assessment Zoning Analysis of loan and mortgage activity Census mapping Zoning area development analysis Development site plan mapping Facility site location Analyze renewal projects Land-use analysis Land value analysis Strategic planning
	Rhinelander (WI)	—	Impact analysis Project planning
Demographic analysis	Los Angeles (CA)	—	Census data analysis
	Champaign (IL)	—	Demographics

Place	Department	Application
East Lansing (MI)	—	Population/density
City of Hopkins (MN)	—	Analysis of population data
		Census data manipulation
Suffolk County (NY)	Department of City Development	Geocoding
		Demography
Cincinnati (OH)	—	Growth forecasting
Dayton (OH)	—	Census block group data
		Usage patterns
Columbia (SC)	—	Demographic analysis
Chattanooga (TN)	—	Census demographics (future)
Austin (TX)	—	Census and planning geographics
		DIME maintenance (future)
Alexandria (VA)	—	Spatial data analysis
Green Bay (WI)	—	Political/demographic analysis
Orange County (FL)	—	Addressing
Schaumburg (IL)	—	Comparative analysis
Indianapolis (IN)	Department of Metropolitan Development	—
General Dutchess County (NY)	—	Major projects review

Administration	City/County	Users	Applications
Equipment maintenance	Green Bay (WI)	—	Equipment reliability analysis
		Utility Department	Equipment performance analysis
			Estimating/equipment maintenance
Information services	Concord (NH)	Municipal	Public information (future)
	Newark (NJ)	Division of Data Processing	Utilities/water and tax information
	Hamilton County (OH)	Regional Computer Center	Land-related information
	Austin (TX)	—	Geocoding
Records management	Alhambra (CA)	Administration	Public complaint tracking/location
	Burbank (CA)	Administration	Inventory analysis
	Long Beach (CA)	Administration	Productivity/inventory
			Project management
	Colorado Springs (CO)	—	Land records management
	Hamilton County (OH)	County Recorders Office	—
	Utah County (UT)	Recorders Office (future)	—
	Madison (WI)	—	Rural land management
			Land records

Finance	City/County	Users	Applications
Budget administration	Kenai (AK)	—	Finance (future)
	Green Bay (WI)	—	Cost of service analysis
			Market analysis
			Finance
Property assessment	Huntsville (AL)	—	Tax neighborhood boundary management
			Sales analysis
	Alameda County (CA)	Assessor's Office	Data management/update
			Locate/inventory/attribute ownership
			Mapping modifications
			Merging school districts
			Ownership transfers
	Alhambra (CA)	—	Assessment information/zoning
	Fresno County (CA)	Assessor's Office	Zoning ordinances
			Permits
	Los Angeles (CA)	City Clerk	Tax/permits
	Redlands (CA)	Assessor's Office	Property ownership
	Brevard County (FL)	Property Appraiser	—
	Gainesville (FL)	Property Appraiser	Zoning
			Housing information
	Hernando County (FL)	Assessor's Office	—
		Appraiser's Office	—
	Orange County (FL)	—	Property appraisals
	McDonough County (IL)	—	Property appraisals
	Boston (MA)	—	Real estate research
	City of Hopkins (MN)	—	Property ownership
			Special assessments

307

Finance

City/County	Users	Applications
St. Paul (MN)	—	Property assessment
		Special assessments
		Land value
		Land ownership
Clark County (NV)	Assessor's Office	Parcels
Broome County (NY)	—	Renter/owner ratios
Riverhead (NY)	Assessor's Office	—
Suffolk County (NY)	Real Estate	Real estate/tax services
	Assessor's Office	Tax/assessment categories
Cincinnati (OH)	—	Real estate research
Cleveland (OH)	—	Real estate research
Pittsburgh (PA)	—	Real estate research
Chattanooga (TN)	—	Identification of unassessed properties (future)
Utah County (UT)	Assessor's Office (future)	—
Arlington (VA)	—	Personal property assessment/registration

	Portsmouth (VA)	Clerk of Courts	Real estate information
	Green Bay (WI)	Public Works Department	Uses/tax assessment
		—	Tax assessment
			Property records
			Depreciation/taxes/inventory
	Milwaukee (WI)	Tax Assessor	Neighborhood assessment
			Location analysis
Revenue collection	Kenai (AK)	Kenai Peninsula Borough	Tax assessment/collection
	Alameda County (CA)	Assessor's Office	Tax remittances/funds dispersing
	Burbank (CA)	—	Business tax/license system (future)
	Huntington Beach (CA)	—	Taxation
	Orange County (FL)	—	Taxation
	Garfield Township (MI)	—	Taxation
	Freeport (NY)	—	Assessment of revenues
	Suffolk County (NY)	Assessment/Taxes	Delinquent taxes
	Dayton (OH)	—	Delinquent real estate taxes
	Milwaukee (WI)	Common Council	Property tax delinquencies

Management	City/County	Users	Applications
Coordination of services	Huntsville (AL)	—	Dispatching assessors
	Alhambra (CA)	Public Services Building and Safety	Maintenance scheduling Analysis of workload
	Los Angeles (CA)	Community Development Department	Production/maintenance
	Denver County (CO)	—	Nine-digit ZIP Code conversion
	Portsmouth (VA)	—	Work order scheduling (future)
Elections	Mesa (AZ)	City Clerk	Voting districts
	Phoenix (AZ)	Salt River Project Administration	Voting system
	Long Beach (CA)	City Clerk	Political divisions
	Los Angeles (CA)	Registrar of Voters	Elections
	San Bernardino (CA)	—	Precinct development
	Town of Hadley (MA)	Board of Elections	Political boundaries
	Suffolk County (NY)	—	Districts
	Arlington (VA)	Election Commission	Voter registration
	Milwaukee (WI)		Redistricting aldermanic district boundaries
Legislative activity	Orange County (FL)	—	Redistricting voting wards
	Oklahoma City (OK)	—	Municipality management-related regulations Intergovernmental services

Application	Location	Department/Office	Use
Long-range planning	Petersburg (FL)	—	Need/evaluation/implementation/management
	Frankfort (KY)	—	Long-range planning
	St. Paul (MN)	—	Management planning
Policy development	Phoenix (AZ)	Maricopa Association of Governments	—
	Silver Spring (MD)	Division of Data Processing	Policy decisions
	Newark (NJ)		Policy decisions
Districting	Mesa (AZ)	Planning and Development Department	Districting
	San Bernardino (CA)	—	Special districts
	San Diego (CA)	Planning	Jurisdictional boundaries (future)
	East Lansing (MI)	—	Administration district/political divisions
	Washoe County (NV)	—	Precincts
	Arlington (VA)	Planning	Districting
Land-records management	Minneapolis (MN)	—	Property management
	Hamilton County (OH)	Recorder's Office/real estate developers	Parcel identifications
General	Pittsburgh (PA)	Planning Department	Vacant lot inventories
	Chattanooga (TN)	—	Land data/property records (future)
	Milwaukee (WI)	City Engineer	Parcel/right-of-way mapping
	Aspen and Pitkin (CO)	—	Decision making
	Silver Spring (MD)	—	Management decisions
	St. Paul (MN)	—	Management decision making

Unknown Applications	City/County	Users	Applications
	Vancouver (Canada)	City of Vancouver	—
	Berkeley (CA)	City of Berkeley	—
	Burbank (CA)	Assistant City Manager	—
	Chino (CA)	City of Chino	—
	Gilroy (CA)	City of Gilroy	—
	Lake County (CA)	—	—
	Marvin County (CA)	—	—
	Sacramento (CA)	City of Sacramento	—
	City of Santa Ana (CA)	—	—
	Stamford (CT)	Communications Center	—
	Davenport (IA)	City of Davenport	—
	New Orleans (LA)	City of New Orleans	—
	Worcester (ME)	City of Worcester	—
	Anne Arundel County (MD)	—	—
	Charlotte (NC)	Charlotte Municipal Information Systems	—
	Murfreesboro (TN)	City of Murfreesboro	—
	Mesquite (TX)	City of Mesquite	—
	Richardson (TX)	City of Richardson	—
	Sandy (UT)	Sandy City	—
	Chesterfield (VA)	Chesterfield County	—
	Pierce County (WA)	Pierce County	—

Glossary

Absolute Map Accuracy The accuracy of a map in relationship to the earth's geoid. The accuracy of locations on a map that are defined relative to the earth's geoid are considered absolute because their positions are global in nature and accurately fix a location that can be referenced to all other locations on the earth. Contrast absolute map accuracy with relative map accuracy.

Application A use of a computerized information system that can be repeated more than once, with different data. An application is one type of use of computers, such as "map updating," "routing vehicles," "assessing property," or "updating property characteristics."

Assessment The establishment of a monetary value of a parcel of land. The assessment value of a parcel of land is established by local government (usually the Tax Assessment Department), which uses a number of factors such as size, use, location, restrictions, and improvements (building characteristics). The assessment value is multiplied by a tax rate to determine the amount of taxes owed by the owner of the parcel. Also known as "assessed value."

Attribute A descriptive characteristic of a feature. An attribute asks a question about it: what, where, how big, how many, when, etc. The answers to the questions are the values stored in a data base. Cartographic attributes describe how to display map information (color, length, height, width, etc.), while nongraphics data attributes describe the mapped feature (what it is, what it cost, when it was built).

Automated Mapping/Facilities Management (AM/FM) A computer graphics system that has data-base management capabilities for at-

313

tribute data associated with features that can be displayed on a map. The focus of an AM/FM system is the storage, retrieval, update, manipulation, and display of map and facility information and is generally used in utility and public works applications. AM/FM systems often lack spatial analysis capabilities.

Backup   A copy of data and computer programs on magnetic tape or disk for safekeeping in the event that the original is destroyed, damaged, or lost.

Base Map   Fundamental map information, either as one layer or as a combination of layers, which is used as a standard framework upon which additional data of a specific nature are overlaid. A base map is used to control all other sources of spatial data and usually includes a geodetic control network as part of its structure. Base maps for urban geographic information systems usually include streets, parcels, and hydrography.

Block   A geographic area of a metropolitan region that is usually bounded by street segments. A block may also be bounded by rivers, railroads, shorelines, political boundaries, or other features that form logical boundaries for a polygon. The block is the smallest geographic unit used by the U.S. Census Bureau for reporting census data and for coding geographic base files such as the DIME and TIGER files.

Block Group   A subdivision of a census tract, also bounded by street segments, natural or man-made features, or political and corporate limits, which combines blocks into contiguous groups having a population of approximately 1000 people.

Buffering   A spatial analysis function that forms a polygon around a point, line, or other polygon by locating its boundaries at a certain distance from the point, line, or polygon. An example is the creation of a polygon whose boundaries are 500 feet from the bank of a river.

Cadastral Maps   Graphical portrayals of the legal descriptions of land parcels. Typically, cadastral maps are tax maps created and maintained by local government for property assessment and taxation purposes. They also include boundaries of easements, public rights-of-way, and some physical features that are necessary for identifying the boundaries of legal interests in land. Sources of information for cadastral maps are legal survey notes.

Cadastre   A record of interests in land, encompassing both the nature and extent of interests. Generally, this means maps and other descriptions of land parcels as well as the identification of who owns certain legal rights to the land (such as ownership, liens, easements, mortgages, and other legal interests). Cadastral information often includes other descriptive information about land parcels.

Cartographic Data   Attributes of cartographic features that are stored in a computer system for display on a map.

Cartographic Entity   A physical object that can be seen on the ground and can be represented on a map.

Cartographic Feature   Something that can be named or assigned an identifier (such as a street, a manhole, a property line, a building, or an intersection) and can be located on a map.

Cartographic Object   The digital record that contains attributes of a cartographic feature and is stored in a computer system.

Cathode Ray Tube (CRT)   A component of a computer terminal that displays data and graphic images on a screen.

Census Tract   A geographic area, bounded by street segments, natural or man-made features, or political and corporate boundaries within a governmental unit. Census tract boundaries were established by local governments for using statistical data provided by the U.S. Census Bureau and generally include areas having a total population of approximately 4000 people.

Central Processing Unit (CPU)   The main processor of a computer that controls the functions it performs.

Choroplethic Map   A thematic map that displays statistical data for geographic areas by filling polygons of the areas with colors or grey tones in accordance with a legend that defines the range of statistical values associated with each particular color or grey tone.

Cluster   A spatial grouping of features on a map. When features are clustered on a map, there is usually some phenomenon causing a relationship among the clustered features (such as incidents of disease, crime, etc.).

Computer-Aided Drafting and Design (CADD)   The use of a computer graphics system to create, modify, manipulate, and display drawings. When used for mapping, CADD systems treat map sheets or drawings as separate entities with little or no continuity across map sheets. While some CADD systems may contain limited data-base management capabilities, they seldom have map registration or map projection transformation capabilities.

Computer-Aided Mapping (CAM)   The use of computer graphics for map design, creation, and maintenance. (See also Computer-Aided Drafting and Design.)

Contour Map   A map that displays lines having the same elevation across the area covered by the map. Topographic maps include elevation contour lines in addition to other planimetric information.

Coordinate Filtering   See Curve Fitting.

Coordinate Geometry (COGO)   Automated mapping software that translates the alphanumeric data associated with a survey (distances, bearings, coordinates, etc.) into digital map information for creating and updating a digital cartographic data base.

Coordinate Transformation   Automated mapping software that mathematically converts coordinates from one frame of reference to coordinates of another frame of reference. Most geographic information systems, for example, can translate latitude and longitude coordinates into state plane coordinates by using coordinate transformation programs.

Copy Parallel   A function of automated mapping software that allows the operator to identify a line on the display of a map and copy that line a specified distance parallel to it, forming two parallel lines.

Cost Justification   The process of gaining approval to invest money in a new venture. The cost justification of a new computer system answers the question: "Is it worth the cost?"

Coterminous   Having the same or coincident boundaries. Two adjacent polygons are coterminous when they share the same boundary (such as a street centerline).

Cursor   A symbol displayed on the screen of a computer terminal, the location of which is controlled by a human operator using the keyboard, a joystick, a tracking ball, or a digitizer. Usually in the form of a cross-hair or blinking square, a cursor is used in a GIS to identify locations or cartographic features displayed on a map for processing by the computer.

Curve Fitting   An automated mapping function that converts a series of short, connected straight lines into smooth curves to represent features that do not have precise mathematical definitions (such as rivers, shorelines, and contour lines). Also called "coordinate filtering."

Data Aggregation   The statistical summarization of data by data item. In a GIS, data aggregation occurs when the data values are summarized by geographic area.

Data Base   A collection of interrelated data that is stored in a computerized information system to serve one or more applications and is independent of the computer programs that use it. A data base can consist of more than one data file.

Data-Base Schema   A logical description of data stored in a data base. The schema not only defines the names of the data items and their sizes and other characteristics, but it also identifies the relationships among the items (all the data associated with a block is also associated with the census tract into which the block falls, for example).

Data-Base Management System (DBMS)   A collection of computer programs that are used to organize and use data stored in a data base. Typical functions of a DBMS include the logical and physical linkage of related data elements, the retrieval and verification of data values, and other data management functions such as security, archiving, and updating.

Data-Based Information System   A computer system containing one or more data bases that are controlled by a data-base management system. It is used to process input transactions (as in a transaction-based computer system) as well as information retrieval and summarization commands. A data-based information system separates the data from the computer programs that process the data.

Data Element The generic name given to a data item. Since different data bases are designed by different computer systems analysts at different times, the names of data items that represent the same entity may differ. A data element is the common reference to these different names, such as an "address," which may be called "location" in one data base and "premise" in another.

Data Field See Data Item.

Data Integration The combination of data bases or data files from different functional units of an organization or from different organizations that collect different data for the same features (such as properties, census tracts, street segments, etc.) This combination of data provides added intelligence to the data that are available on geographic features.

Data Item The smallest unit of information that has meaning when describing an entity. A data item is also referred to as a "data field," which is a physical storage location on a record in a data base or file.

Data Modeling The identification of the data needed to perform the essential functions of an organization. Data modeling is a method for designing computerized information systems based upon the data that must be available in a data base and can be available for computer programs to retrieve, manipulate, update, store, and display in order to satisfy the information requirements of one or more functions of the organization.

Datum An ellipsoid used to represent the surface of the earth mathematically so that coordinates can be assigned to locations on the surface. See Local Datum, Global Datum, and North American Datum.

Deed A written instrument that, when executed and delivered, conveys an estate in real property or interest therein. An essential component of a deed is a description of the property ("legal description") that accurately and precisely locates the property without oral testimony. Land surveyors translate this description into land subdivision plats, assessors plats, cemetery plats, certified survey maps for public recordation, and plats of other surveys for filing with the County Surveyor. These plats (or the legal descriptions in the deeds) form the basis for locating property lines in a GIS.

Densification The process of adding monuments to a survey control network for the purpose of simplifying a survey that requires a monument as a reference point for recording the location of some feature. A survey can be simplified by having a control network monument nearby, because time and risk of error is minimized.

Development Cycle   The series of tasks that must be performed in order to render an information system operational. Typical tasks include: problem definition, analysis of needs, feasibility study, cost/benefit study, systems design, systems development (programming), testing, training, and implementation.

Digital Cartographic Data Base   The attributes of map features stored in a computer system.

Digital Map   The representation of cartographic features in a form that allows the values of their attributes to be stored, manipulated, and output by a computer system. A digital map is a data base or file that becomes a map when a GIS produces a hardcopy or screen display output.

Digitize   The act of converting a map into digital form by encoding the spatial coordinates of features on the map with the aid of a handheld electronic device called a Digitizer.

Digitizer   A small device guided by a human operator over the surface of a digitizing table to either position the cursor on the computer terminal screen or to identify the locations of cartographic features on a hardcopy map for the purpose of entering their coordinates into the computer. (Also called "mouse" or "puck.") The word is also used to refer to the person controlling the device.

Dot-Density Map   A thematic map that displays the occurrence of particular entities at specific points on a map. Dot-density maps display the spatial clustering of features having common characteristics.

Easement   The legal right to use land that is not owned by the person or legal entity holding that right. A utility easement allows a utility company to dig the ground on land it does not own for the purpose of accessing its underground facilities.

Edge matching   A function of automated mapping software that manipulates features at the edge of adjacent map sheets to fit them into continuous features that cross map sheet boundaries.

Edit   The act of correcting erroneous data or updating information in a computer system. After editing a map in a GIS, its cartographic features are changed. After editing a nongraphics data base, the values of attributes stored in the data base are changed.

Ellipsoid   A surface all plane sections of which are ellipses or circles. In a GIS, an ellipsoid refers to a three-dimensional shape whose surface can be mathematically defined so that coordinates can be assigned to locations on the surface. An ellipsoid that most accurately represents the surface of the earth is called a Datum.

Facility   A specific component of the urban infrastructure such as a water main, street pavement, telephone pole, or street light.

Flat File   A structure for storing data in a computer system in which each record in the file has the same data items, or fields. Usually, one field is designated as a "key" that is used by computer programs for locating a particular record or set of records or for sorting the entire file in a particular order.

Flow Analysis   A spatial analysis function that processes attributes of line segment data about the transportation of liquids along a network (such as volume, rate of flow, etc.).

Font   The style of typeface used to display text on a map produced by a GIS.

Geocode   An identifier assigned to both a map feature and to a data record containing attributes that describe the entity represented by the map feature. Common geocodes include: addresses, census tracts, and political and administrative districts. Geocodes are also referred to as "location identifiers."

Geocoding   The process of assigning geocodes to data describing entities that can be located on a map. The most common geocoding function in local government is the use of a geographic base file such as the DIME File to assign census tract codes or other local geocodes to data records containing addresses.

Geodetic Survey Network   A survey network referenced to the earth's geoid. Measurements taken from monuments tied to a geodetic survey network accurately place features on the earth so that they are precisely located relative to other features anywhere else on the earth. Contrast a geodetic survey network with a Local Survey Network.

Geographic Base File (GBF)   A digital record of map information. This means that a GBF can contain the cartographic data necessary for displaying a map, but it also means that a GBF can contain only the attributes of the geography represented on a map (such as address ranges of a street segment, intersections of streets, and boundaries of geographic areas). The most common GBFs are the DIME Files and the TIGER Files. (Also known ᷉ "geobased file.")

Geographic Information System (GIS)   A computerized data-base system for capture, storage, retrieval, analysis, and display of spatial data.

Geoid   The shape of the earth as a three-dimensional spheroid that coincides with the surface of the earth at sea level and extends in an imaginary surface through the continents with a direction of gravity that is perpendicular at every point. Although the geoid appears to be a sphere, it is actually an ellipsoid flattened at the poles and bulging at the equator. This makes it impossible to represent the geoid mathematically. When a mathematical representation of the surface of the earth is needed (such as in the assignment of coordinates to positions on the earth), then a reference ellipsoid that most closely resembles the surface of the earth is used (since an ellipsoid can be represented mathematically). A reference ellipsoid that most closely represents the entire geoid is called a Global Datum, while a reference ellipsoid that most closely represents the geoid in a certain geographic area of the earth is called a Local Datum.

Geoprocessing   The manipulation of data based upon its geographic nature. In addition to the manipulation of maps and cartographic data bases, geoprocessing also involves the manipulation of nongraphics data in data files whose records contain geocodes.

Global Datum   A datum that most closely represents the earth's geoid for the entire surface of the earth. Contrast with Local Datum.

Global Positioning System (GPS)   A method used in surveying that uses a constellation of satellites orbiting the earth at very high altitudes. GPS technology allows accurate geodetic surveys by using specially designed receivers that, when positioned at a point on the earth, measure the distance from that point to three or more orbiting satellites. Through the geometric calculations of triangulation, the coordinates of the point on the surface of the earth are determined.

Hardcopy   One of many forms of output from a computer system. A hardcopy output is the presentation of information on paper or other physical media that can be viewed by the human eye. A hardcopy map is produced in a GIS by plotting its points, lines, symbols, and text strings onto a physical medium.

Hierarchical File Structure   A structure for storing data in a data-base management system in which there is more than one type of record in the data base. A hierarchical data structure is also called a "tree structure" because some records are subordinate to others in a "one-to-many" relationship (also called a "parent–child" relationship). Access to a record is made first through its parent record.

Horizontal Data Integration   The combining of different data bases containing attributes of the same entities, but for varying purposes by different functional units of an organization or different organizations. A property tax assessment file, for example, consists of parcel and building characteristics used in the tax assessment function of local government. A building permit file, containing information about building permits that describe the same entities (buildings), is used in the permit processing function of government. When data about each building permit are related to the assessment data for each associated property in the tax assessment file, the data are considered to be integrated horizontally.

Hydrography   The description of rivers, lakes, oceans, and other water-related features.

Information System   A computer system in which the data stored will be used in spontaneous ways that are not fully predictable in advance for obtaining information. See Data-Based Information System.

Infrastructure   A system of facilities such as streets, sewers, water mains, gas mains, telephone lines, and other permanent physical installations that are the basic support systems for living in an urban environment.

Justification The position (left, right, center) of the value of a data attribute relative to some predetermined location. On a map, a text string or symbol is "justified" relative to the position at which it was digitized. In a nongraphics data-base record, the value of an attribute is "justified" relative to the beginning position of the data field containing the value.

Key A data item used to identify or locate a record in a data file.

Land Information System (LIS) A computerized data-base management system, often a geographic information system, that contains cadastral, land use, natural resource, environmental impact, and other data describing land.

Land Title Recordation System The process of recording information about legal claims to land, including transfer of ownership based upon deeds and wills; security interests based upon judgments, mortgages, and construction liens; and other interests such as easements and public and utility rights-of-way.

Large-Scale Map A map displayed for a small geographic area (ranging from a street intersection to a few square miles) and displayed at a scale with a large numeric ratio, such as 1 inch equals 400 feet or greater (since the ratio 1 inch equals 400 feet is greater than 1 inch equals 24,000 feet, it is considered a larger scale). A map drawn at a scale of 1 inch equals 400 feet or greater is usually considered a large-scale map.

Local Datum A datum that most closely represents the surface of the earth (the earth's geoid) in a particular geographic area of the earth (such as North America or a particular state or county).

Local Survey Network A survey network based upon monuments not located in relation to their absolute position on the earth's geoid. Maps that are created from measurements based upon a local survey network are accurate only within the geographic area covered by the monuments and are not accurate in relationship to the earth's geoid. An example of a local survey network is the subdivision platting process based upon the Public Land Survey System. In many cases, the plats are tied to PLSS monuments at section and quarter-section corners that have not been surveyed to determine their geodetic coordinates. Contrast a local survey network with a Geodetic Survey Network.

Location Identifier See Geocode.

Map Accuracy The conformity between the location of a feature displayed on a map and its true location on the earth relative to an accepted control network or datum. Accuracy is expressed in terms of an error, in linear units. Horizontal accuracy refers to the position of a feature on the face of the earth. Vertical accuracy refers to the height of a feature relative to sea level.

Map Layer   A grouping of homogeneous map information (such as curb lines) that is stored or identified separate from other map layers. A map layer can be analyzed, manipulated, and displayed individually or in combination with other map layers. A map layer is also known as a "map level" or a "map coverage."

Map Projection   The transformation of a map from a sphere to a plane surface. Because the geometric and mathematical relationships of angles, areas, distances, and directions are difficult to maintain in a map projection, there are an infinite number of ways to transform a spherical surface to a plane. Only a few map projections, however, are in common use: Lambert's conformal conic, Transverse Mercator, and stereographic.

Map Scale   The mathematical relationship between distances on a map and actual distances on the earth, expressed as a ratio of inches (or centimeters) on the map to feet (or meters) on the earth. Urban maps usually are drawn at scales ranging from 1 inch equals 10 feet to 1 inch equals 200 feet. In a GIS, map information is stored in the computer without any scale because a particular scale is assigned only when the map is actually displayed.

Metes and Bounds   A method for surveying land by recording the locations of features by their distance and angle from points whose locations are recorded and can be found on the earth.

Monument   A physical marker embedded in the earth in as permanent a form as possible. The location of a monument is recorded for future use when measuring the location of an entity on the ground.

Multipurpose Cadastre   A parcel-based land information system consisting of a geodetic reference frame, a base map that uses the geodetic reference frame for control, and a cadastral overlay that is controlled by references to both the geodetic reference frame and features on the base map. Unique parcel identifiers on the cadastral overlay allow integration of the data with other parcel-based legal, fiscal, and resource data.

Nearest Neighbor   A spatial analysis function that uses proximal analysis to identify a feature or group of features that are closest to a feature of interest on a map.

Network Analysis   A spatial analysis function that uses the topological structure of lines to follow a path along an interconnected network and then process attribute data associated with the line segments. See Flow Analysis, Routing, and Optimum Path Analysis.

Network File Structure   A structure for storing data in a data-base management system in which there is more than one type of record in the data base. A network file structure provides a "many-to-many" relationship in which a "child" record can have more than one "parent" record. A network file structure is also known as a "plex structure."

Normalization The decomposition of a complex data structure into simple, flat files (relations). Normalization creates separate files that have common data fields, replacing the associations represented by pointers and keys in hierarchical and network data structures.

North American Datum (NAD) A local datum that provides the most accurate mathematical representation of the surface of the earth (the earth's geoid) for locations in North America. The most recent North American Datum that has been identified is the North American Datum of 1983 (NAD 83). The North American Datum used prior to 1983 is the North American Datum of 1927 (NAD 27).

On-Line An adjective describing a computer system in which the input data enter the computer directly from their point of origin and output data are transmitted directly to where they are used, usually at a computer terminal. On-line computer systems allow the operator to interact with the data so that inputs do not have to be grouped into batches, processed by programs activated by someone else, and printed at a central location.

Optimum Path Analysis A spatial analysis function using network analysis to determine the path along a network of lines that meet some predetermined criteria (such as shortest length, shortest travel time, etc.).

Output The result of processing data in a computer program. Output can be in the form of a tabular listing of data or summary statistics, a map, or other data in hardcopy, screen display, or optical disk, as well as a computer file in hard disk, floppy disk, or tape form.

Parcel Also known as a "land parcel," a parcel is a physical area of land that can be defined as a polygon and has legally defined boundaries that are recorded by local government in a Land Title Recordation System. A land parcel is owned by one person or other legal entity (such as a corporation, partnership, etc.).

Planimetric Map Graphical representation of the physical features of land and other physical entities located on the land. Planimetric maps are created from field surveys or aerial photographs because they identify the shapes and locations of entities that can be seen. Topographic maps are planimetric maps that also identify the elevation of the land in a series of contour lines.

Plotter A hardware device for a computer system that produces hardcopy graphic outputs called "plots." Most plotters use either pens for producing the final output or an electrostatic process for transferring ink to paper or other hardcopy medium.

Point-in-Polygon A spatial analysis function that identifies whether points are inside a polygon boundary or outside a polygon boundary. The attributes of those points identified as being within the polygon are then processed for analysis by displaying symbols on a map, computing statistics of the attribute values, listing attribute values in tabular reports, etc.

Pointer   The location of one record in another related record of a data base that allows a computer program to access the data in the latter record when the former record is found. A pointer "points" a computer program to another record that contains data related to the initial record accessed.

Polygon Fill   A function that shades, colors, or otherwise patterns the area inside the boundaries of a polygon. In a geographic information system, polygon fill functions are used to create choroplethic maps.

Polygon Overlay   A spatial analysis function that uses Boolean logic (AND, OR, NOT, etc.) to create new polygons from the intersection of the boundaries of polygons from two different maps. An example is the overlay of flood plain districts onto parcel maps to create new polygons that are portions of parcels within a flood plain.

Polygon Processing   A spatial analysis function that uses the boundaries of polygons to identify which points, lines, and areas within the polygon to select for further processing. See Point-in-Polygon, Polygon Overlay, and Choroplethic Map.

Polygonization   A spatial analysis function that creates polygons from the attributes of features on a map based upon certain spatial or statistical criteria. See Spatial Aggregation and Buffering.

Property   The combination of land parcels and improvements (buildings) owned by a person or a legal entity. While a parcel defines an area of land, property also includes all physical entities located on the parcel that have a monetary value. In common usage, the distinction between a parcel of land and a property is negligible.

Proximal Analysis   A spatial analysis function that displays symbols on a map of location-related features and then identifies which other features are closest to them. See Nearest Neighbor and Dot-Density Map.

Public Land Survey System (PLSS)   A rectangular survey system that utilizes 6-mile-square townships as its basic survey unit. The location of townships is controlled by baselines and meridians running parallel to latitude and longitude lines. Townships are defined by range lines running parallel (north–south) to meridians and township lines running parallel (east–west) to baselines. The PLSS was established in the U.S. by the Land Ordinance of 1785.

Raster   A method for displaying or storing graphic data that uses individual points for processing. On displaying or plotting the data, each point is either displayed (with or without color) or not displayed, depending upon whether it is part of the image or not. On storing and manipulating raster data, each point contains attribute values that are used by computer programs for processing. A raster can be thought of as a grid of points (such as the lights of an athletic scoreboard) that are individually processed to manipulate or display data that represent a larger image or feature.

Record   A logical and, sometimes, physical grouping of related data fields that are treated as a single unit by a computer program. A tax assessment record contains all of the pertinent data needed to calculate the assessed value of a particular parcel of land.

Relation   A flat file that can be combined with another relation in a Relational Data Base system to form a new relation of combined data from the two files.

Relational Data Base   A data-base structure composed of more than one flat file that can be easily combined because of relations between the data in the records. Data in a relational data base are stored as two-dimensional arrays (relations) that can be combined to form new relations for processing by computer programs. Contrast with hierarchical and network data-base structures.

Relative Map Accuracy   The accuracy of a map in relationship to a local survey network that is not tied to the earth's geoid. The accuracy of locations on a map defined relative to a local survey network is considered relative because the positions are accurate only within a certain geographic area covered by the network. Maps from surveys based upon a local survey network are not easily registered to other maps based on the earth's geoid.

Remonumentation   The process of constructing a permanent marker in the ground to replace one that has been destroyed since it was originally surveyed and constructed. The process involves a survey to locate the original point at which the monument was constructed from historical survey records.

Resource Allocation   The assignment or distribution of people, equipment, or money to activities of an organization. In local government this usually means distributing or assigning these resources to different geographic areas of the jurisdiction in order to provide some public service: fire or police response, snow removal, inspection services, health and welfare services, etc. One of the applications of a GIS in local government is to assist managers allocate resources to geographic areas so that public needs within those areas are met with the appropriate amount of public resources.

Routing   A spatial analysis function that uses network analysis to identify a path along a network, usually for moving a vehicle from one point on the network to another point (such as a fire truck to a fire).

Scale   The mathematical relationship between the size of a feature on a map and the actual size of its entity in the real world, expressed as a ratio of map size to actual size. For example, a map scale of 1:100 ("one to one hundred") means that a measurement of one inch on a map represents one hundred feet on the ground.

Small-Scale Map   A map displayed for a large geographic area such as an entire city, county, state, or country. The scale of such a map is considered small because the numeric value of the ratio defined by the scale (say, 1 inch equals 24,000 feet) is small compared with that of a large-scale map is drawn at a larger scale (say, 1 inch equals 50 feet).

Snapping   A function of automated mapping software that makes two lines meet mathematically during the digitizing process. This is used to ensure connectivity between lines when it cannot be verified visually.

Spatial Aggregation   A spatial analysis function that creates polygons whose boundaries are established from the clustering of features having similar attribute values.

Spatial Query   A function of a geographic information system that allows a user to find and display a map or attributes of features located on a map.

Spatial Reference Framework   A coordinate system based upon survey measurements and used to define the relative and/or absolute locations of features in a geographic information system.

Street Segment   The physical entity of a travelway that begins at one street intersection and ends at the next. It is the basic unit for building topology in a street network and is defined uniquely by the name of the street and the range of addresses along the segment.

Survey Network   A series of physical monuments placed on the earth and used by land surveyors as reference points to locate features recorded on maps and other drawings or survey records. The locations of these monuments may be referenced globally on the earth's geoid (as in a geodetic survey network) or only within a small geographic area (as in a local survey network).

Tabular Listing   An output of a computer system, either in hardcopy or screen display form, that presents data in rows and columns.

Terminal   A hardware device of a computer system, usually consisting of a keyboard and a cathode ray tube (CRT) screen, used to communicate with the computer. A graphics terminal contains a high-resolution CRT that provides sharp lines that appear to be smooth when displayed.

Text (or Text String)   A combination of letters, numbers, or words, usually referred to in a GIS when describing information displayed on a map output from the system.

Text Placement   A function of automated mapping software that places a text string at a specific location on a map.

Thematic Map   A map that displays data related to a specific topic such as property assessments, housing quality, crime, health problems, etc. Data can be represented on a thematic map in a variety of manners: dot-density, choroplethic, contour, etc. A base map may be used as an overlay to provide geographic reference to the location of the data.

Topographic Map   Graphical representations of the locations of natural and man-made features on the earth. Topographic maps are planimetric maps that also include elevation contours of the land.

Topological Data Structure   The explicit definition of how map features represented by points, lines, and areas are related. For example, a street segment is explicitly defined as having two endpoints at each of its intersections with other street segments. These endpoints are, in turn, explicitly defined as endpoints of other street segments. All street segments are defined as being either a border of an area (such as a block), or entirely within an area (such as a ZIP Code area). Attribute data stored within the data bases of a GIS usually define these explicit relationships.

Vertical Data Integration   The aggregation of detailed data from a data base into consecutively larger groupings for use by higher organizational levels in management and policy applications, usually in the form of summary or statistical reports or data files. Crime data, for example, are often summarized by squad area for analyzing workload by squad. These data are also further summarized across all squad areas in each police district for establishing resource allocation policies. Crime data are also summarized by census tract and then integrated with population data by census tract to calculate crime rates based upon population for comparing geographic areas or trends over periods of time.

Work Station   A combination of hardware and software normally used by one person to interact with a computer system. A GIS work station consists of a keyboard, a graphics terminal, a digitizing table or tablet, a digitizer, and (optionally) a plotter. A work station can also consist of a central processing unit (CPU), which allows it to operate as its own computer in a network of other computers. These "intelligent work stations" reduce the demands on the central computer and allow the transfer of programs and data files between computers.

Windowing   A feature of automated mapping software that allows an operator to identify a smaller geographic area than that which is displayed on a screen and display it on a larger scale on the same screen or an adjacent screen. Also called "zooming."

Workload Balancing   The process of defining work assignments such that each worker has the same amount of work to accomplish. When work is assigned by geographic area (such as police squad area, tax assessment district, or public health nursing district), a geographic information system can be used to identify the locations of the work with respect to the boundaries of the assignment area. This allows managers to adjust their boundaries so that there is a balance in the amount of work among the different areas.

Zooming   See Windowing.

Index